Territorial Imaginaries

T0391829

Territorial Imaginaries

Beyond the Sovereign Map

Edited by Kären Wigen

THE UNIVERSITY OF CHICAGO PRESS

Chicago & London

The University of Chicago Press, Chicago 60637

The University of Chicago Press, Ltd., London

Published 2025

Printed in Canada

34 33 32 31 30 29 28 27 26 25   1 2 3 4 5

ISBN-13: 978-0-226-83900-4 (cloth)

ISBN-13: 978-0-226-83901-1 (e-book)

DOI: https://doi.org/10.7208/chicago/9780226839011.001.0001

Library of Congress Cataloging-in-Publication Data

Names: Wigen, Kären, 1958– editor.

Title: Territorial imaginaries : beyond the sovereign map / edited by Kären Wigen.

Description: Chicago : The University of Chicago Press, 2025. | Includes bibliographical references and index.

Identifiers: LCCN 2024033897 | ISBN 9780226839004 (cloth) | ISBN 9780226839011 (ebook)

Subjects: LCSH: Cartography. | Sovereignty. | Human territoriality. | Boundaries. | Geographical perception.

Classification: LCC GA101.5 .T47 2025 | DDC 304.2/3—dc23/eng/20240812

LC record available at https://lccn.loc.gov/2024033897

♾ This paper meets the requirements of ANSI/NISO Z39.48-1992 (Permanence of Paper).

For Martin

CONTENTS

FOREWORD

The present volume grew out of a two-day conference at Stanford University, held in May 2022, under the rubric of "Re-mapping Sovereignty: Representing Geopolitical Complexity." The vision behind this event was to bring together a dozen historically minded scholars who were either making incisive critiques of conventional models of territoriality or pointing the way toward new accounts and representations. When initially contacting them, the conference organizers, Kären Wigen and Martin Lewis, held out the enticement of meeting at Stanford University's David Rumsey Map Center, where participants would be able to engage with cartographic imagery on wall-size, high-resolution screens. The maps on view would be of two kinds: original works (including treasures from the Rumsey collection) and retrospective visualizations, designed to convey the contours of polities past. Together, we hoped, such images would help us better grasp the overlapping, ambiguous, and contested nature of territoriality through time.

The goals of gathering at the Rumsey Center were twofold. First, we sought to create an opportunity for conversation among far-flung thinkers who were independently making important advances on remapping sovereignty, but who were not necessarily aware of one another's work. To that end, we reached out to scholars in a range of fields, including geography, history, art history, government, law, and anthropology. We also strove for regional diversity, including specialists on East Asia, Central Asia, the Middle East, Europe, and the Americas. By mixing up the disciplines, and by putting scholars of Asia in dialogue with those who study the Euro-Atlantic world, we hoped to spark new insights into the distinctive shapes of power—and the challenges of mapping them accurately—across a wide swath of time and space.

The second goal was to use those conversations as a springboard to launch a comparative volume, which you now hold in your hands. As discussed in more detail in the introduction, the book traces an arc from the archives to theory to the cartography lab. Part I offers a series of case studies from outside the time space of modern conventional cartography; here, historians of four distinct world regions reveal the knotty nexus of maps and territorial claims in their respective parts of the world. The implicit (and sometimes explicit) critique of the Westphalian model that emerges from these textured studies is given a more abstract turn in part II, whose authors offer warnings about the dangers of conventional political mapping as well as models for alternative modes of representing states. Finally, the essays in part III take an experimental tack, showing promising paths forward through counter-cartography, map art, and digital design.

As it happened, two dramatic events shaped the context in which the conference ultimately took place. One was Russia's invasion of Ukraine—a shock to the international system that made headlines throughout the spring of 2022. Even as participants were putting the finishing touches on their essays, the world's attention was riveted by a violent assault on national sovereignty. For those attuned to cartography, this was also a resonant moment for thinking about the challenge of rendering the complexities of territoriality on maps. No longer was sovereignty or its representation an abstract issue for scholars to debate; overnight, it had been transformed into a life-and-death matter that dominated the mediascape.

The other inescapable feature of the global environment that spring was the COVID-19 pandemic. When invitations were issued in 2021, there was no way to know if the contributors would be able to gather in person. This made planning tricky, and the suspense continued down to the wire. In the end, we were lucky: Bay Area COVID-19 restrictions began to lift in April 2022, making "Re-mapping Sovereignty" one of the first in-person gatherings permitted on the Stanford campus in two years. Members of the Stanford community and the general public were able to join the conference at the Rumsey Center for two days of stimulating talks.

None of this would have been possible without the support of many people. At Stanford, we are indebted to our generous sponsors: the School of Humanities and Sciences, the Department of History, and especially the staff of the David Rumsey Map Center. Without their encouragement, skills, and funding, the conference simply could not have taken place. Director Salim Mohammed and his talented team enabled a technologically exacting hybrid event to unroll seamlessly. They also designed and mounted a beautiful exhibit featuring many of the maps that eventually made their way into this book—an exhibit that curious readers can visit in its digital form on the center's website.

We are also grateful to those who lent their social skills and intellectual acumen both to the conference gathering and to shepherding the manuscript that resulted from it. Abby Smith Rumsey, Marie Price, and William Rankin were sharp interlocutors who drew out themes and connections between papers. Yuki Hoshino was unsurpassed as a resourceful and efficient conference assistant. Jay and Nancy Hamilton graciously hosted the conference dinner. Sayoko Sakakibara assisted with the many figures and permissions. Finally, Susannah Engstrom at the University of Chicago Press deftly guided the book through the review process and into production, and Johanna Rosenbohm provided expert copyediting.

Thanks to the support of these generous people and institutions, this stimulating collection of essays has now reached your hands. While the contributors have no illusions that ours will be the last word on the subject of territorial imaginaries, we are excited to share this work and look forward to continuing the conversation.

Stanford University
March 2024

INTRODUCTION

Kären Wigen

> We live enmeshed in thick webs of borders and boundaries. Most are out of sight and conscious awareness, yet they all impinge in some way on our lives as integral parts of our real and imagined geographies and biographies.
> —EDWARD SOJA, "Borders Unbound," from *B/ordering Space*

In 2013, a Stanford freshman dreamed of making a map. Assigned to contribute to a campus counter-atlas—that is, an atlas the administration would not or could not make—she wanted to wind back time and show the university lands as they might have looked in Ohlone days. Her starting points were seemingly basic questions: Where had Native inhabitants lived? Where were their hamlets, their hunting grounds, their burial sites? What shape did their territories assume? No Indigenous maps of the area survived, and early European maps of California made only crude gestures, if any, toward Native peoples' presence. So Jen West sought out the university archaeologist to request a map of relevant sites on the sprawling campus. The interview was granted; the map was not. In the interest of protecting Native remains, the university declined to share their location. West would have to deduce where the Ohlone had lived by learning what they valued.

Initially, the prospects looked good. Ohlone lifeways had revolved around rivers, obsidian quarries, hunting grounds, and vistas of Mount Diablo (the cosmic axis of the Ohlone world)—sites whose locations could be inferred from a topographical map. But while resources and ritual sites could be deduced, territorial markers were stubbornly elusive. The density of Native settlement in the Bay Area implied that the local landscape had once been rich in placenames. Yet preconquest toponyms—the most basic expressions of human territoriality and communal history[1]—had left few traces here. More thoroughly than in most of North America, Native polities in California had been swept off the landscape, overwritten by the fencelines and surnames of latecomers like Leland Stanford.

Figure 0.1 shows how West handled this problematic lacuna. Unable to supply Indigenous place names, she underlined their erasure: heavy black lines stood in where toponyms should have been. And with no way to reconstruct Ohlone territory, she simply cut her base map at the jagged property lines of the university's 8,180-acre parcel while letting woodlands and rivers spill out to the paper's edge. The resulting geobody was as strange to her peers as it would have been to the Ohlone. The sandstone quad that anchors most campus maps was barely visible, shrunk to a few centimeters and displaced to a corner. Nor was that the only cause of estrangement. Evoking long-ago lifeways in a high-tech idiom, this crowded canvas had an eerie effect. Its blacked-out labels, in particular—incongruous among the bright polygons and leaping stags—stirred a sense of disquiet that lingered in the mind.

That lingering disquiet was what brought *Ohlone Stanford Lands* to memory ten years later, when I sat down to pull together this collection. In her own thoughtful way, I realized, Jen West had flagged the four major themes that inform the present volume.

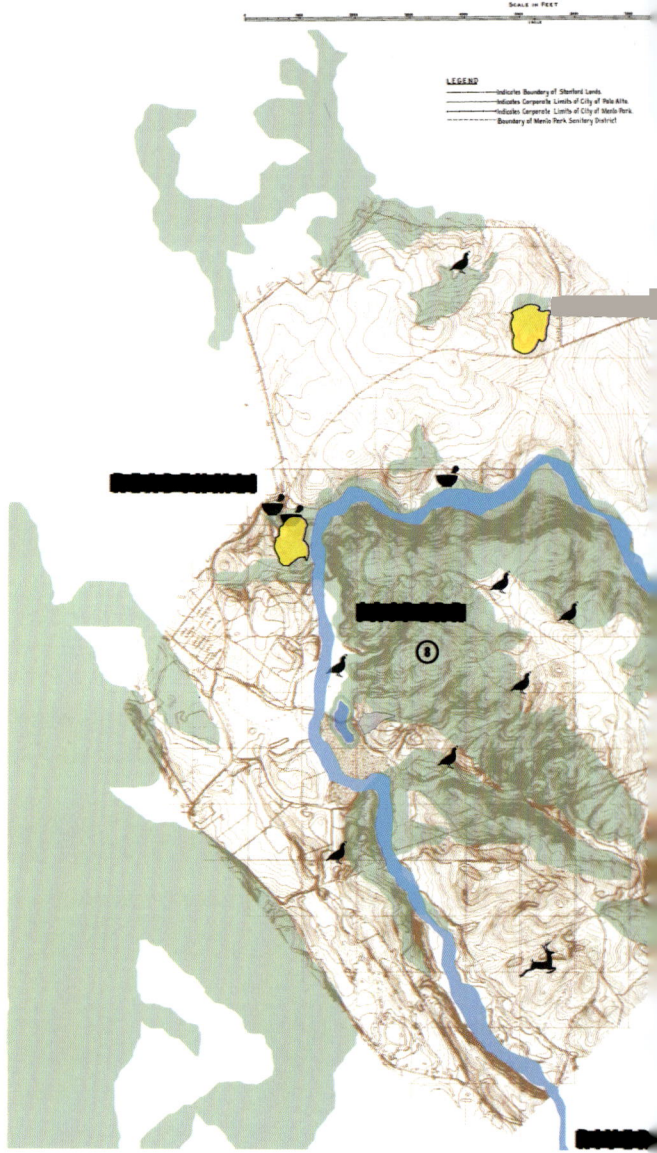

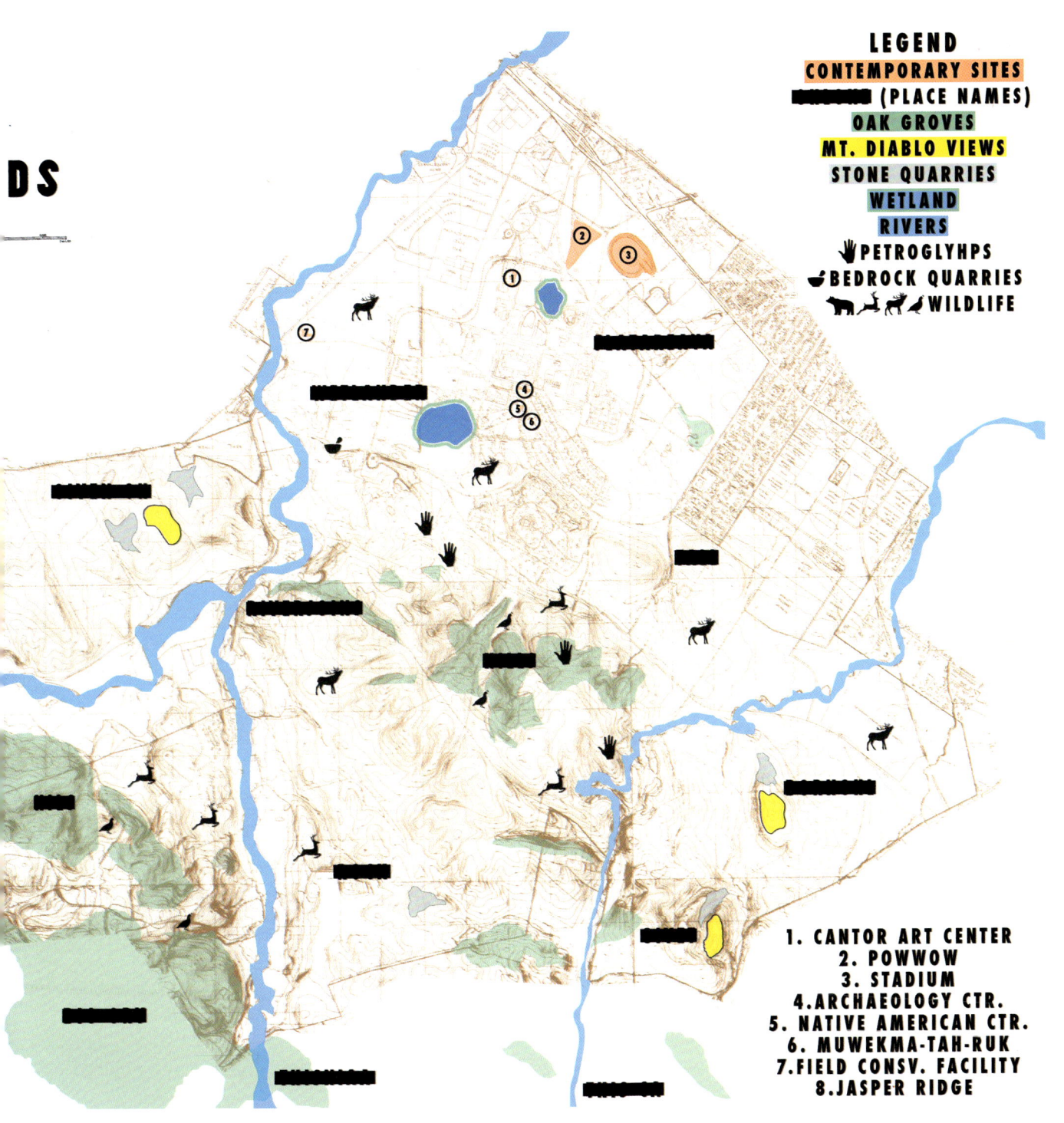

LEGEND

CONTEMPORARY SITES
▬▬▬▬ **(PLACE NAMES)**
OAK GROVES
MT. DIABLO VIEWS
STONE QUARRIES
WETLAND
RIVERS
✋ **PETROGLYHPS**
⛏ **BEDROCK QUARRIES**
🐻🦌🐦 **WILDLIFE**

1. **CANTOR ART CENTER**
2. **POWWOW**
3. **STADIUM**
4. **ARCHAEOLOGY CTR.**
5. **NATIVE AMERICAN CTR.**
6. **MUWEKMA-TAH-RUK**
7. **FIELD CONSV. FACILITY**
8. **JASPER RIDGE**

FIGURE 0.1. Jen Ward West, ~~Ohlone~~ *Stanford Lands*. Map from a final project for History 95N, Maps and the Modern Imagination, fall 2013, printed in *Stanford Counter-atlas 2013*, ed. Kären Wigen. Courtesy of Jen Ward West.

Foremost among them is the one that West found most problematic: territoriality. As the ten chapters in this collection bear witness, territoriality has taken many forms. Biologists are not alone in recording a host of behaviors by which "one or more individuals actively defend a home range against other members of their own species" (to cite a trans-species definition of the term);[2] archives and ethnography add many more. With humans as with our other animal kin, those behaviors sometimes take benign and even beautiful forms. A turf claim might be broadcast through plumage displays or trumpeting calls, folksong or dance. But when strut and song are challenged, tooth and claw come out.

Not surprisingly, those who theorize territoriality have written at length about those teeth and claws. Violence is embedded in our definition of political space; for Max Weber, the state was precisely "that human community which (successfully) lays claim to the *monopoly of legitimate physical violence* within a certain territory."[3] This volume, too, takes violence seriously. Militarized borders and interstate conflicts figure one way or another in every chapter. Nonetheless, our emphasis lies elsewhere; the second term in our title highlights not the fact of aggression but acts of imagination.

To speak of territoriality and imagination in the same breath might seem naïve in a world where neighboring states are yet again at each other's throats. Still, we are convinced these terms belong together. No contributor to this project would suggest that territorial claims rest solely on ideas; but if ideas are not the whole story, neither are arms. What

we find particularly engaging, and in need of fresh eyes, are the intricate arrangements through which territoriality has played out over time—and the wide array of rituals, symbols, and surrogates that have been devised to represent them. To explore that dual diversity means taking a long look at what, following John Gillis, we might call "territories of the mind."[4]

The project of exploring geographies of the mind, or spatial imaginaries, has gained traction in the humanities of late. Through close readings of art, maps, travel writing, textbooks, fiction, and more, geographers and historians have joined visual and literary scholars to illuminate the enormous range of ways that human societies have envisioned their own and others' worlds. Such studies have uncovered fascinating fields of discourse at every scale, from the micro to the meta.[5] Together, they have done much to illuminate how spatial ideas are integral to place making and world building.

Applying that approach to territoriality is the core of the present project. To do that well, we believe, means wrestling with two additional issues: sovereignty and the map. These, too, are essential keywords for this project. Because territory is understood as *controlled* space—that is, as space that is not only visited or occupied but governed and defended—coming to terms with territoriality inevitably means thinking about sovereignty. On the one hand, sovereignty is inherently territorial; the sovereign state is defined as one that overrides all other powers that may cross its jurisdiction, while brooking no outside interference in its internal affairs. By the same token, territory is an expression of sovereignty. The *Encyclopedia Britannica* tellingly defines *territoriality* as "the monopolization of space by an individual or group."

But is a genuine monopoly on space ever possible? Even in theory, the several attributes of sovereignty do not necessarily coincide. The ability of a government to exercise power in its own lands is one thing; the power to exclude external actors from its structures of authority is another.[6] And when it comes to the actual exercise of power over terrain, cracks between the *de jure* and the *de facto* appear at every turn. The more one looks, the more one sees. In fact, *sovereignty* may be the most fraught term in the geopolitical lexicon, at once "amorphous, elusive, and polysemic."[7] As Wendy Brown concludes, "We have known all along that sovereignty has been, if not a fiction, something of an abstraction with a tenuous bearing on political reality."[8] The historian James Sheehan concurs. "As a doctrine," he writes, "sovereignty is usually regarded as unified and inseparable; as an activity, however, it is plural and divisible."[9]

That recognition is fundamental to this book. The scholars who signed onto this venture did so out of a shared interest in probing territorial anomalies. Sovereign control, they knew, was *not* always projected, much less practiced, as unified or inseparable. Over most of the world—until quite recent times—plurality and divisibility have been less the exception than the norm. That point is underscored in one after another of the case studies that follow. Nor should it surprise us. For until the twentieth century, the world abounded in empires ruled by dynastic monarchies: assemblages of peoples linked together through war or dowry under one crown, yet retaining varying degrees of autonomy in law and governance. Patchwork sovereignty of this kind persisted in Europe long after the famous Peace of Westphalia in 1648—the conventional birthdate of the modern state system. For another two centuries and more, political space in Central Europe's Holy Roman Empire remained fractured and layered, marked by "the omnipresence of condominiums and exclaves, the limitations of the rulers' territorial superiority, [and] the frequency of overlapping and contradictory political claims."[10] Rather than being based on contiguous, cohesive units, Luca Scholz argues, "the polities of the early modern Empire are better understood as a system of channels, corridors and checkpoints unevenly distributed in space."[11] This territorial system persisted until 1806.

The Holy Roman Empire may have had an unusually complex geography for Europe, but it was hardly alone in the world. East Asian sovereignty was equally intricate. The Manchu Qing ruled over a vast composite empire (with separate accommodations for their Tibetan, Mongolian, Manchu, and Han domains, as well as for tributary states like Korea); the Japanese archipelago sustained similar complexity on a smaller scale. In the words of Fabian Drixler, William Fleming, and Robert Wheeler, the Tokugawa realm before 1868 was "a collection of some 250 states and statelets," including domains whose holdings were "no more than portfolios of widely scattered villages."[12] One of the larger domains, the fiefdom of Satsuma, was itself a mini empire, ruling Ryukyu as a subfief. And Ryukyu in turn was even more complex, at once a Janus-faced polity (paying nominal obeisance to both Beijing and Edo) and a hierarchical one (with Okinawa dominating the smaller islands in the Ryukyuan archipelago).[13] In such a landscape, sovereignty could be neither exercised nor represented in jigsaw-puzzle style. Even basic boundaries were unclear. Ronald Toby memorably described the Tokugawa regime as having a "ragged edge;"[14] more recently, Mark Ravina has shown that internal boundaries, too, were often left vague, long after Japanese cartographers had the tools to nail them down if they wished. Border ambiguity, for this compound state, was a deliberate choice.[15] To do justice to this kind of geopolitical complexity requires envisioning sovereignty in fresh ways.

Which brings us to our final critical term: the map. Among the many signs and surrogates that have been devised to represent territorial claims, none may be more fraught than the political map. By design, such maps confer coherence and stability on their subjects.[16] For Denis Wood, this is in fact their most important function: "It has been essential that states appear as facts of nature, as real enduring things, things like mountains; and at all costs to obscure their recent origins . . . and their tenuous holds on tomorrow."[17] Little wonder, then, that conventional cartography has limited resources for conveying the more intricate ways that terrain has historically been shared and contested. Some such arrangements are challenging to describe in words, much less in maps—for, as Lauren Benton vividly conjures it, the premodern world was shot through with spatial and legal anomalies. "Empires did not

cover space evenly," she writes, "but composed a fabric that was full of holes, stitched together out of pieces, a tangle of strings."[18] Even countries like the United States, founded in the spirit of the Enlightenment, were riddled with compromises over territorial control. Forging thirteen separate colonies into a coherent country was such a prolonged and uneven process that not until after the Civil War did the country come to be referred to in the singular, turning the once-common phrase *the United States are* into antiquated grammar.[19] And even then, the Union remained fragile and perforated.[20]

Textbook maps cannot readily represent these complexities, for conventional cartography follows a Westphalian playbook. It treats the globe as a game board, neatly divided between a discrete set of players (notionally sovereign states) that may vary widely in size but are otherwise of the same categorical type. Specialists know better; scholars routinely lament that the part of the world they know best deserves more sophisticated treatment. Those who commission maps share a growing awareness of a fundamental contradiction between the simplicity of our mapping toolkit on the one hand, and the contortions and contingencies of territoriality in particular places on the other. Doing justice to the empirical record with the tools at hand has proven frustrating. Conventional cartography simply cannot help us think outside the Westphalian box.

The result is a widening search for fresh approaches. As William Rankin writes, "There's no way to use jigsaw-puzzle maps to challenge the core assumption of the jurisdictional model: that space is composed of homogeneous blocks with sharp boundaries. We can't just zoom in, we can't just find the 'right' boundaries, and we can't easily switch to using data organized in some other way. It's a closed system." Instead, Rankin is taking what he sees as "the only way out": rejecting the basic premise and drawing the world differently.[21] Digital humanists around the country are responding to the same pull. From Stanford to Austin to New Haven and beyond, spatial history laboratories have begun experimenting with alternative ways to visualize how sovereignty actually works.[22]

In its own way, ~~Ohlone~~ *Stanford Lands* contributed to that effort. Running up against the limits of conventional cartography, Jen West chose not to dodge those limitations but to expose them. Her map draws attention to the lacunae in our knowledge, as well as to the arbitrariness of our boundaries. Given the constraints under which she worked, West could only gesture toward the need to fundamentally reimagine territoriality. But we can push the project further. For scholars who can access them, archives the world over hold rich resources for this work.

Overview of This Volume

The present volume takes up where West's map left off. Part I offers a series of case studies that explore mapping practices before the nation-state. Ranging far in space and time, these early chapters help us tease out alternative modes of imagining territoriality before modern mapping conventions took hold. Valerie Kivelson starts off by introducing a pioneering atlas of Siberia that was produced at the turn of the eighteenth century for Peter the Great. A royal commission might have been expected to yield maps that flattered the tsar, with sweeping territorial claims, bold political borders, and cartouches of the kind produced for early modern courts everywhere from Iberia to Scandinavia. Instead, what Kivelson finds in Remezov's atlas is an improvised cartographic idiom whose focus lay elsewhere altogether. What concerned this Siberian resident was ethnography, mobility, and the fur trade, not geopolitical borders. His maps offer a provocative counterpoint to conventional representations of territoriality under an absolutist ruler.

From Russia we move to Inner Asia, where the Mongolian scholar Lhamsuren Munkh-Erdene explores nomadic sovereignty both before and after the Qing conquest of the steppe. Munkh-Erdene's chapter mounts a forceful argument against the lingering view that nomads had no concept of territoriality. He begins by tackling the tenacious idea that sovereignty in human societies has gone through a universal phase shift, from power over people to rule over territory. Charting the rise of this people-to-territory idea in European political theory, he shows both the untenability of such an account for Europe and its pernicious effects when projected onto the rest of the world. Moving to Mongolia in particular, he points to passages from *The Secret History of the Mongols* that spell out the Chinggisid theory of rule—one where control of people and control of territory were inextricably linked. He goes on to argue that the 1640 Great Code of the Mongols went further than the current UN charter in guaranteeing the territorial rights of lesser powers within a multilayered interstate system.

The third chapter takes us into the Chinese imperial archive. Here, the historian Peter Bol juxtaposes two pairs of maps, at radically different scales, that would seem to make an incongruous set. Bol's first images, carved onto a famous stone stele in the twelfth century, conjure a vast terrain of time and space—the canvas on which a frontier upstart proposed to stake a new dynasty.[23] The second pair of images, from a seventeenth-century gazetteer, are smaller in every way. Humble in medium and modest in scope, these black-and-white woodcuts depict a single township in terms of its resident lineages. Despite the differences, Bol sees a similar contrariness at work in both sets of images. Whether expansive or intimate, they were made against the imperial grain. Their implicit visions of sovereignty cut against the conventions of their day; each projected a realm that transcended or defied the official spatial imaginary of the administrative, centralized state. And while both experiments proved short lived (earning from Bol the designation "maps for failed states"), their survival gives us tools with which to think beyond the official spatial imaginary of their day and ours.

The last chapter in part I looks at a striking case of a *refusal* to map. The historian Ali Yaycıoğlu focuses here on a watershed moment: the Karlowitz negotiations of 1699, when the Ottomans were purportedly brought into the Westphalian system. Picking up on an element that most historians have overlooked, Yaycıoğlu focuses on the losing side's refusal to endorse a map as part of the peace talks. As he recounts, a conflict broke out at the treaty table between the Austrian and Turkish diplomats over how to describe the new border between their respective domains. While the Habsburgs wanted to draw up the final borders in Karlowitz—based on their generals' knowledge of military positions at the moment when hostilities ceased—the Ottomans balked. To do such a thing would be premature, their negotiator insisted; a durable border could be worked out only by consulting local actors on the ground, and ensuring that each village retained access to the resources needed to sustain life. Effectively, Yaycıoğlu argues, this debate pitted generals' knowledge against local knowledge—and local knowledge won out.

These wide-ranging non-Westphalian moments help lay the empirical groundwork for the arguments in part II, "Pushing Back against the Sovereign Map." Each chapter in this section offers a critique of conventional cartographic practice. Chapter 5 contrasts two contemporary border wars where maps played different roles. Coauthors Alexander Murphy and Cy Abbott start by posing a comparative question: Why have some territorial conflicts been more amenable to settlement than others? Their answer is that border wars are particularly hard to resolve when national identities are rooted in ethno-nationalism. The authors develop that argument by contrasting two twentieth-century disputes over territory: one between Greece and Turkey, which smolders to this day, and the other between Ecuador and Peru, which has been largely resolved. The stubbornness of the

Greece–Turkey divide, they suggest, is rooted in the mutually exclusive territorial imaginaries of Greece and Turkey, grounded in deep-seated, ethnocultural narratives. Ecuador and Peru present a very different case. Despite decades of armed conflict, the long-running border war between these South American neighbors was for all intents and purposes put to rest in 1998. The key, Murphy and Abbott say, lies in each region's history of national formation. Unlike Greece and Turkey, Ecuador and Peru were characterized from the start by a mixture of peoples. In this context, ethnocultural conceptions of the national space simply never gained ground—reducing the emotional freight of the region's long-running border wars.

Next, Jordan Branch contributes a thought experiment that argues for verbal rather than visual modeling of sovereignty. His provocative essay proposes that political theorists should reconceive states as composite entities made up of ideas, infrastructures, and representations: ideas about organization, authority, and action; infrastructures of control and communication; and representations in visual, linguistic, and other forms. In this view, mapping is given an integral role in state formation and persistence even as the use of cartographic illustration is deliberately eschewed in favor of formulating the model in the abstract. Where Branch himself has previously argued that the territoriality of the modern state was shaped and made possible by developments in mapping technology, here he moves from the idea that "maps *made* the state" to the possibility that "mapping *is* (part of) the state." Likewise, he suggests

that perhaps the state persists not only *because of* the hegemony of the political map of the world, but *in* that representation itself.

The final essay in part II is a cautionary critique by the anthropologist Franck Billé, who bids us beware the map. Decades of research into territoriality and its expressions in East Asia have left Billé leery. Staid though they may seem, images of the national geobody carry a powerful psychic charge. Like images of our physiological selves, they have come to form an integral part of the proprio-sensory apparatus that orients modern humans in our environment. After probing how these emotionally laden symbols stir us (for better and worse), Billé ends his chapter by scanning the horizon for alternatives. For representing the nation, he comes down on the side of noncartographic expressions: words or symbols that can evoke a sense of belonging without all the hazards of maps.

While these critiques make a vital contribution to a book that aims to push beyond the sovereign map, they do not get the last word. Instead, we move in part III from critique to counter-cartography. The chapters in this closing section point to possible ways forward. Each highlights ongoing experimentation, whether in the artist's studio or the cartographic laboratory.

Part III opens with a meditation from the art historian Barbara Mundy, who casts her eye back to the fateful Spanish conquest of the Aztec Empire. First, through a close reading of a 1524 map of Tenochtitlan that emerged from that conquest, she shows that the cementing of the association between map and sovereignty in the early sixteenth century had as its aim and consequence the displacement of Indigenous peoples. For the former Aztec lands, that displacement would play out in two acts; if native Mexicans were first dispossessed as sovereign rulers, they would later be dislocated as contemporaneous peoples. Both tropes left traces all over cartography. Whether taking the form of anachronic renderings of Mexico City itself or of timeless Indigenous figures in the foregrounds of maps, the depiction of Indigenous people as out of time continued down to the nineteenth century. But Mundy does not leave off with this bleak observation. Instead, she fast-forwards to contemporary studios and museums, showcasing living Mexican and Chicana artists who are talking back to this tradition. By incorporating and interrogating some of the most iconic elements of early modern Mexican cartography (including the very 1524 map of Tenochtitlan) into their work—and by using figuration to bring Amerindians back into the cartographic space—these artists are profoundly challenging established relationships between maps, Indigenous people, and state power.

Geographer Guntram Herb takes us next to the Weimar Republic, illuminating the nationalistic map campaign that arose when German cartographers honed their craft to reclaim their country's lost territories after World War I. Herb's essay begins by laying out the insights of critical map studies concerning the power of maps, their relationship to persuasion, and their emergence as evidence for national territorial claims. He then brings these insights to bear on the Weimar campaign: comparing map designs and messages, analyzing knowledge networks, and assessing the campaign's overall impact. Herb sees important lessons in this case for our collective project of rethinking territoriality. By their very design, he finds—particularly the use of areal shading to identify and distinguish areas of ethno-national majorities—interwar maps worked to efface ethnic diversity within the lands where German speakers lived. In so doing, these maps provided a veneer of scientific legitimacy for the disastrous expansionism of the 1930s and '40s. Yet maps of this kind are not always tied to hegemony. In a brief coda, Herb shows how similar techniques can be deployed in the interest of destabilization and critique.

A related set of experiments is underway in cartography labs across the world. As Luca Scholz reports in the final chapter, both digital and analog mapmakers are finding novel ways to represent the ambiguities and subtleties of how territoriality has historically been practiced. Scholz was trained as a historian of the Holy Roman Empire, a complex assemblage that has traditionally been portrayed as an array of contiguous and self-contained if fragmented polities. But recent scholarship, including his own, has challenged that characterization at every point, and he and others are now actively experimenting with tools to better visualize the intersecting, discontinuous, power-sharing mechanisms of the empire's spatio-political order. In particular, Scholz highlights three concrete ways that nonexclusive forms of dominion can be represented on maps: by reframing density as dilution; by emphasizing the infrastructure of connection (roads) over that of separation (borders); and by acknowledging diachronic modes of power sharing.

With this, the present volume's arc—from case study to critique to counter-cartography—reaches its end. But the project of reimagining territory is far from finished. Even as this volume goes to press, new scholarship is documenting the alarming extent to which the global order has become perforated by a myriad zones of sovereign exception. From military bases to foreign-leased territories, offshore tax havens, free trade zones, and more, the planet now hosts a swarm of compact areas where the allegedly normal rules don't apply. Invisible on a world map, these sites are not mere relics of the premodern spatial order. On the contrary, such sovereign "anomalies" have proliferated in recent decades until they now number in the thousands. As Daniel Immerwahr writes, special zones "have grown from a marginal to a defining feature of the economy."[24] This should sound a wake-up call; the collective challenge of reimagining and remapping sovereign territory has never been more urgent. Along with the contributors to this venture, I look forward to continuing the conversation—and to as-yet-unimagined breakthroughs in the maps and models to come.

PART I

Mapping Practices
before the Nation-State

1

Valerie Kivelson

Ambiguous Territories

Mapping Siberia in the Era of Peter the Great

Maps can be full of surprises, often startling us by upsetting our expectations. Maps of the eastern reaches of the Russian Empire from the era of Peter the Great surprise in both what they depict and what they ignore. Petrine maps reveal a complex and unexpected approach to boundaries—one that does not conform to standard expectations of what constitutes a sovereign state. Most definitions of sovereignty insist that it functions within a clearly delimited territory, where its claims can be asserted unambiguously and its jurisdiction enforced. Early eighteenth-century Russian cartography showcases a world of difference and a different world, where the Westphalian framing so central to our usual understanding of sovereignty turns out to be of little relevance. So different was this conceptual mapping of borders that it may allay the concern raised by Franck Billé in his essay in this collection, that we are trapped by a "political cartography, [by] a neat arrangement of discrete entities with no gaps or overlaps," that we lack the "ability to imagine other forms of spatial belonging." It may even answer Jordan Branch's call (also in this volume) for a different kind of mapping, a new way to conceptualize the state "in a world 'seduced' by the nation-state map."

Petrine maps of Siberia reflect a very different way of imagining territorial sovereignty. They convey an alternative approach to territory and statehood. But, having raised such an exciting prospect, I must immediately dash any hopes that this particular way of breaking out of the Westphalian vise might lead to liberatory alternatives. Especially at the current moment, working under the dark shadow of Russia's brutal violation of Ukrainian sovereignty, it is important to acknowledge that the different forms of sovereign mapping evident in these maps are not one whit more gentle or humane.

Peter the Great, who reigned from 1682 until his death in 1725, energetically promoted expeditions to map his vast Siberian holdings and, notably, to figure out where they were and what resources they contained. As Russian forces spread across Eurasia in the sixteenth, seventeenth, and eighteenth centuries, their representatives, with no deep tradition of mapmaking to draw on, had to work out formulas for conquest and rule on the ground.[1] The resulting

I want to thank James Meador and Kären Wigen for their many excellent suggestions.

maps reflected a geopolitical world unconcerned with the notion of sovereign rule over a contiguous, integral space. Peter's mapmakers, in their outlines, symbols, and labels, presented a world where sovereignty was spotty and pocked, enforced or waived in specific places, with specific groups or individuals. It was relational and negotiated. Ambiguity was most marked where Russians eventually recognized that they had bumped up against other constituted polities, like China or Bukhara; but it punctuated Russia's mastery within its tenuous borders as well.

The obvious place to begin a study of cartography and territorial imagination is with Semen Ul'ianovich Remezov's spectacular so-called "ethnographic map" of Siberia (figure 1.1) Remezov (1642–after 1720)

made his mark as the foremost cartographer of Siberia, and left the largest corpus of maps of Siberia from the late seventeenth and early eighteenth century. By the order of his sovereign, this talented polymath from the western Siberian capital of Tobolsk undertook the task of mapping all of Siberia. He fulfilled the commission with a glorious, oversized manuscript atlas, the *Chertezhnaia kniga*, or Sketchbook, that he completed between 1698 and 1701 and presented to the Siberian Chancellery in Moscow. The atlas includes twenty-one maps, each introduced by a short title page or lengthier explanatory text.

In addition to that masterpiece, Remezov worked on two other versions, also each preserved in a single manuscript: the *Khorograficheskaia chertezhnaia kniga*, or Chorographic Sketchbook (1697–1711); and the *Sluzhebnaia chertezhnaia kniga*, or Working Sketchbook (1697–1720).[2] The Chorographic Sketchbook is a small, fat notebook of horizontal sketch maps (*chertezhi*), depicting segments of Siberia's main rivers. It charts their tributaries, sources, and outlets, as well as the human settlements and natural landmarks that surrounded them. The Working Sketchbook is perhaps the most intriguing of the lot. Where the *Chertezhnaia kniga* was a polished set of maps submitted to the authorities and the Chorographic Sketchbook is evidently Remezov's own collection of rough drafts, the glorious Working Sketchbook contains the most ornate, polished versions of his maps, as well as beautiful, precise copies of European maps, elaborate baroque reveries, architectural drawings, and technical plans for brickworks that he established near Tobolsk. What the book is, then—his own working notebook or a formal work to be submitted to the state—remains unclear.[3] But since it contains numerous maps of Siberia, it tends to be counted as the third of his great manuscript atlases.

Remezov drew on the cartographic efforts and geographic knowledge of many other people, so through him we can see a varied community of explorers and mapmakers. He acknowledged the contributions of these other explorers and mapmakers, whose maps he copied or adapted, and he also noted that he drew on information provided by local residents, both Russian and Indigenous. Increasingly accurate maps of Kamchatka show his commitment to incorporating geographic information as soon as it reached him.[4] So, constrained though we are primarily to the work of one individual, the source base is not as paltry as it seems.

The ethnographic map serves as the grand finale of Remezov's *Chertezhnaia kniga*. The ethnographic title attached to the map centuries later; he called the work *Sketch of the Boundaries and Locations of All Siberia, the City of Tobolsk, and All the Various Cities and Settlements and Steppes*. A. V. Psianchin remarks on the map's uniqueness as the first and "practically the only" ethnographic map made prior to the beginning of the nineteenth century.[5] Oriented with south at the top, the map shows the various "lands" that occupy the space between "the tsardom of Great Muscovy"—the large orange splotch on the right margin—and the speckled black and gray of the Arctic and Pacific Oceans, at the bottom and along the left margin. Each colored blob, Remezov informs us, represents the territory of a particular group of people, or *rod*—a word that forms the root of *birth*, *kind*, or *lineage*. Land and people are united in these sharply delineated spaces. The striking use of color underscores divisions. Remezov assures the reader that the various groups never stray across their borders; if they do, they risk provoking warfare. He was not, then, indifferent to boundaries. But he showed no interest in distinguishing different genres or tiers of boundaries.

In his maps, Remezov played with ideas of territory, boundaries, borders, and sovereignty, developing useful solutions without relying on any theoretical models or abstraction. Boundaries—of regions, of plowed fields, and of what we might recognize as ethnicities—feature prominently. Borders between realms, by contrast, are either indistinguishable from ethnic or tribal divisions or are left completely unmarked. The emphatic dark wavy line that encircles the orange semicircle of the Land of Great Muscovy, for instance, represents the Volga River—not a political boundary at all. The curvy ribbons of white and yellow indicate mountain ranges. A series of polities frame the Russian holdings in the south and east (moving right to left along the top margin: the Dominion of Khiva, the Bukharan Tsardom, the Kazakh Horde, the Yellow, Black, and White Mongols, the Land of the Tsardom of China, Korea, the Giliak Tsardom . . .), but they are distinguished from their neighbors by nothing more than their labels and the same sharp color contrasts that characterize each of the regions. Remezov creates a visual uniformity among the constituent parts, conveying a sense that each of these units—Chukotka, the land of the Ostiak Horde, or the lands of China or Muscovy—was essentially equivalent. The colors, though varied and strongly contrasting, belong to the same rich, autumnal palette. The patches are of roughly equivalent size and irregular shape, labeled in roughly the same way.

Only one element, the large white patch near the center, stands out among the units. It is labeled "the Land of Great Tartaria," and "the glorious city of Tobolsk."[6] The visual and verbal exaltations of Tobolsk reflect Remezov's own passionate (and over-the-top) enthusiasm for his hometown. With the same ardor, on other maps he surrounded the city with radiant red lines, and in his rhapsodic writings he celebrated it as "a glorious angel," protected by God, bringing Orthodox Christian enlightenment to the benighted

steppe. He even mapped this angel, in words, onto the walls, streets, and structures of the city and, on a grander scale, onto the full Siberian expanse.[7] His emphasis on Tobolsk, then, has little to do with its standing as regional capital or its administrative pride of place among other territorial subdivisions. Rather, Remezov drew on a vision of the world deeply informed by attachment to his natal place and by his commitment to a world where Orthodoxy resided in places. Marking contending sovereign territories, indicating political equivalence or hierarchy, were not the intent of the map.

Yet, at the same time, Remezov was alert to questions of dominion. His labels scrupulously indicate which peoples and spaces accepted Muscovite dominion and which did not. Groups that had "come under the tsar's mighty hand," acknowledged his sovereign authority, and agreed to pay tribute (*iasak*, paid mostly in fur) won the approving label of "*mirnye*" (peaceful) or "*iasashnye*" (tribute paying). The holdouts were noted as "*nemirnye*" or "*neiasashnye*" (unpeaceful, non–tribute paying). In this map, only one area displays the "unpeaceful" label: the land of the unpeaceful Samoyeds, shown in the golden-ocher patch third from the left at the bottom, along the Arctic Ocean. Other, more obliging Samoyeds occupy their own regions further west, also abutting the Arctic coast.

The same presence/absence of borders characterizes the rest of Remezov's maps and those of his predecessors and contemporaries. The same casual indifference to state borders pairs with close attention to more granular territorial divisions. We will proceed with an examination of each of these aspects in turn, beginning with the treatment of what we might call international or external borders and then shifting inward, to the kinds of borders and human distinctions that interested the mapmakers more.

The concept of formal international boundaries was not by any means unknown in Petrine Russia. European ambassadors had complained for centuries about the way they were kept waiting at the Muscovite border before being allowed to cross, so their hosts clearly were aware of those dividing lines and their practical and ritual potential. Treaties with Poland set formal divides between the two states, with the Dnipro/Dnieper River serving that role from the mid-seventeenth to late eighteenth century. In the east, however, borders were generally less clear. The degree of uncertainty can be striking. According to John H. Appleby, Peter the Great himself told Edmond Halley "that the River Ob in West Siberia formed 'the boundary of his dominions towards the Sea.'"[8] Unfortunately, Appleby provides no source for this delightfully ridiculous quote, which would truncate Peter's holdings in spectacular fashion, and I haven't been able to confirm it.

International borders do appear on some Petrine maps—for instance, on some versions of the Godunov map of Tobolsk and all of Siberia (figure 1.2). The original of this map, now lost, was commissioned in 1667 by Petr Godunov, then governor of Tobolsk. It was widely copied, circulated, and updated for the next six decades. One of several versions included in Remezov's Working Sketchbook indicates some kinds of interstate borders with dotted lines. Dots

separate the "Chinese Land" from . . . well, something. The most likely contender for the region divided from China is "the Siberian Land," which is prominently labeled but not in a way that sets it apart as that-which-is-not-China. The map, we should note, was made a decade or more after the 1689 Treaty of Nerchinsk set the borders of the Russian and Qing Empires; yet those borders are not evident.[9]

In this map, the metropole plays no role. The Tsardom of Muscovy does not receive even a glancing reference; the term *Russia* is equally absent. If the map is meant to document zones of territorial sovereignty, it seems to do so for everyone but the tsar. The city of Tobolsk radiates a red glow. Tangut, Mungal, Saiantsy, Kirgiz, Kontashin, Ablaev, Kalmyk, Bukharan, Kuchkov, Nogai, and Bashkir territories are all circumscribed by dotted lines, while the Siberian lands lie unclaimed, unlabeled in terms of political affiliation. Yuri Slezkine memorably observed that ethnic Russians were the only nationality within the "communal apartment" of the USSR that lacked their own room.[10] The same might be said about Russians in these maps. Is this a case where Russian is an unmarked category, presumed and taken for granted? Does its absence on the map suggest swaggering power rather than ambiguity? Perhaps; but the acknowledgment in this and other maps of zones of indeterminacy and incomplete or absent control argues against that proposition.

Despite the formality of the 1689 Treaty of Nerchinsk, the map suggests, claims to sovereignty were not imagined territorially. Here, where we might expect the clearest expression of bounded

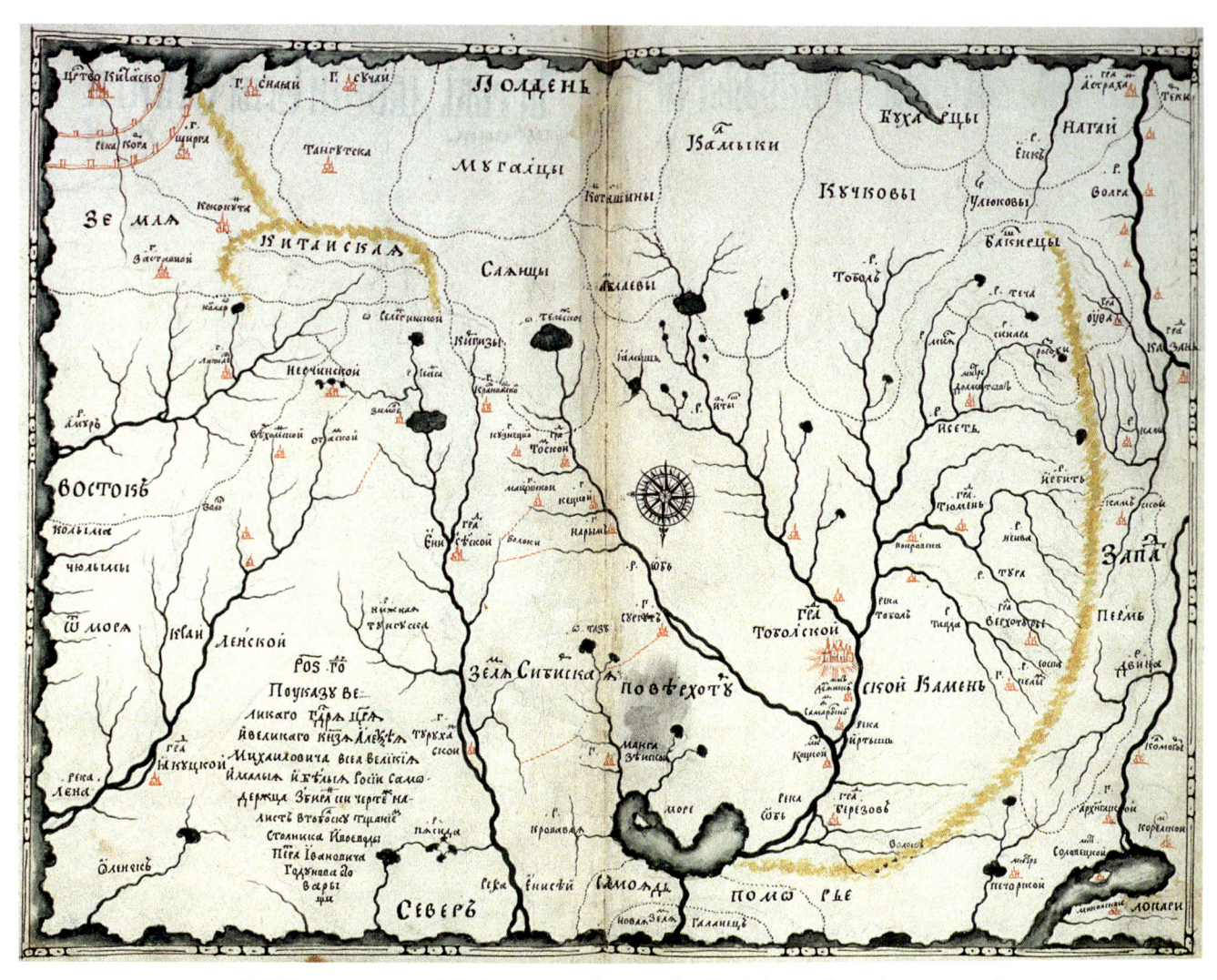

FIGURE 1.2. S. U. Remezov, copy of Godunov map of 1667, in his *Sluzhebnaia chertezhnaia kniga*, (1697–1720), 40 × 61 cm. Rossiiskaia natsional'naia biblioteka, St Petersburg, Ermitazhnoe sobranie, no. 237, fol. 31. Public domain. Available at https://viewer.rusneb.ru/ru/000200_000018_NLR_%D0%9E%D0%A0%2040AC6CFA%20D8B3%20 47B2%20B472%207ABF833BF9D4?page=34.

sovereignty—that is, at the literal border—the concept faces still more serious challenges. The area along the Amur River (the official boundary between the two states by the time Remezov made the map) is marked off with dotted lines, suggesting not a sharp divide but a more indefinite intermediacy. It bears no helpful labeling aside from the reminder that the left side of the map denotes the "East."

A recent, exceptionally vibrant line of inquiry reminds us of the dangers of projecting backward a fictive drama where two expansionist empires met, competed, battled, and negotiated to set a firm line between them.[11] While we know these two empires would eventually divvy up northeast Asia, participants at the time didn't know that. As James Meador writes, Muscovy and the Qing did meet in the Amur basin, but "they did so retrospectively."[12] At the time of their initial encounters, both empires were feeling their way in a complex political landscape where multiple powers—many of whom have vanished from historical memory—participated and competed for resources. Meador writes, with some humor, "It took almost twenty years for Muscovy and the Qing to both realize they were at war. . . . Latency in recognition was a powerful obstacle."[13] The two powers were engaged in a frontier war before either realized it was happening, or even that they shared a frontier. Only when their would-be tribute payers complained of being forced to pay tribute to two sets of intruders did they recognize the problem.[14] Eric Widmer makes a similar point: from the perspective of the Qing and Russian administration, "steppe and forest lands receded from the center of the empire; their job was to define them in terms of the center. Neither was intellectually armed to suspect that these lands, getting further away from one metropolis, might be getting nearer to another."[15] With these caveats in mind, we can try to set aside the expectation that from the start, two vying imperial powers—or, most centrally in this study, the Russian Empire—fought over establishing a boundary line in a disputed frontier region. Instead, we can use the ambiguities to pry apart the very expectation that we should find a standard assortment of cartographic claims to territory. What the maps show instead is a fine-grained, localized, even individualized approach in Russia's quest for something like sovereignty at the edges.[16]

Other Russian maps of the Chinese borderlands adopt variously nebulous approaches to the border.

Remezov occasionally indicated guard posts, but rarely between Qing and Russian holdings. A looseleaf page stuck into the Chorographic Sketchbook, not in Remezov's hand but perhaps made by his son, notes somewhere near where the border should have been: "Here stands a guard," but with no label indicating whose interest this vigilant individual was meant to guard.[17] A map from the *Chertezhnaia kniga* displays a stub of a wall at "the border of Korea with the Chinese (*grani Koren'ia s kitaitsy*)" and notes "a guard—a Chinese man (*karaul chelovek kitai*)" safeguarding the wall. In this same sketch map and in most of Remezov's maps of southeastern Siberia, the Great Wall features prominently; but visually, the maps communicate porousness and ease of travel in and out of China. Red dotted lines wend their way through open gates in the Great Wall, suggesting smooth passage.[18] This may represent an actual moment of openness that followed the 1689 agreement—a period during which trade caravans moved across the forming border and the Orthodox mission in Beijing was established by mutual agreement. Cross-border movement proved short lived. It was truncated by the Treaty of Kiakhta in 1727, when the border hardened and transit was sharply restricted. Toward the bottom of this map, the river Gorbitsa is labeled with the notation "and on it is the border with the Chinese (*a na nei s kitaitsy granitsa*)." The Treaty of Nerchinsk named the confluence of the Gorbitsa and the Shilka Rivers as defining one leg of the Russo–Chinese border, so Remezov got this right. The looseleaf insert of a section of the Amur in the Chorographic Sketchbook again marks the Gorbitsa as the border, this time labeled more actively: "River Gorbitsa. Here the border was set (*tut postavlena granitsa*)."

Intriguingly, the next page in the sketchbook features a pictorial symbol that resembles a chess pawn—a conic figure topped by a tiny ball—labeled "border of China (*gran' kitaiskaia*)"[19] (figure 1.3). Whether or not an actual stone marked the border turns out to be a hotly debated topic. The multilingual scholar Kicengge writes of the "Nerchinsk border stone," that appears on a Manchu "Map of the Amur Region." "On the eastern bank of the river Gerbici, there is a picture of the boundary-stone . . . , the existence of which has been long disputed. An attached legend written in Manchu says *oros emgi hešen be faksalame ilibuha wehe bei*— 'Monument commemorating the treaty concluded with the Russians.'" Kicengge finds, however, that the stone was never set in place. Such a stone was indeed commissioned "in the third month of the 29th year of Kangxi" by the Board of Works. "The order contained instructions concerning the size and proportions of the tablet, inscriptions in five languages (Manchu, Russian, Latin, Mongol, Chinese) and other details," but no suitable stone could be found, and the project was apparently scrapped. Nonetheless, the Manchu mapmaker chose to place it on the map, and Remezov did likewise. Perhaps the stone was indeed erected as planned, or at least these two unconnected mapmakers thought it had been.[20] With the signing of the Treaty of Kiakhta in 1727, more boundary markers were set and transborder passage was curtailed, signaling the first steps

toward consolidating a very new, sharply territorialized conception of sovereignty.[21]

So, borders between empires registered unevenly and inconsistently on Russian maps in this period, mostly remaining invisible and unremarked, though occasionally indicated textually or with subtle visual cues. To be fair, national boundaries did not appear reliably in European mapping either. According to Peter Sahlins, "Until the very end of the seventeenth century, maps generally failed to distinguish provincial and state boundaries, portraying them indistinctly with dotted or dashed lines. But mountains often doubled as political boundaries, and it was not unknown for publishers to highlight in color the mountain ranges that served to separate different territories."[22]

By Remezov's time, however, boundaries of state entities and political units were standard on European maps, and the territorial extent of sovereign authority was usually clearly addressed both visually and textually. European cartographers hired by Peter the Great to map his Siberian holdings put this preoccupation into practice. With text, title, and pictorial vignettes, they made plain who held sway. For instance, Nicolas de Fer titled his 1722 map *Les Etats du Czar ou Empereur des Russes en Europe et en Asie*.[23] Despite the emphatic label, the Siberian landmass is subdivided into a variety of units some within Russia's domain and others not. It is still hard to find imperial borders in de Fer's rendition. An important map that circulated in print in Europe was made by the Swedish Philip Johan von Strahlenberg. Strahlenberg spent ten years in Tobolsk as a prisoner of war, and then stayed on to take part in

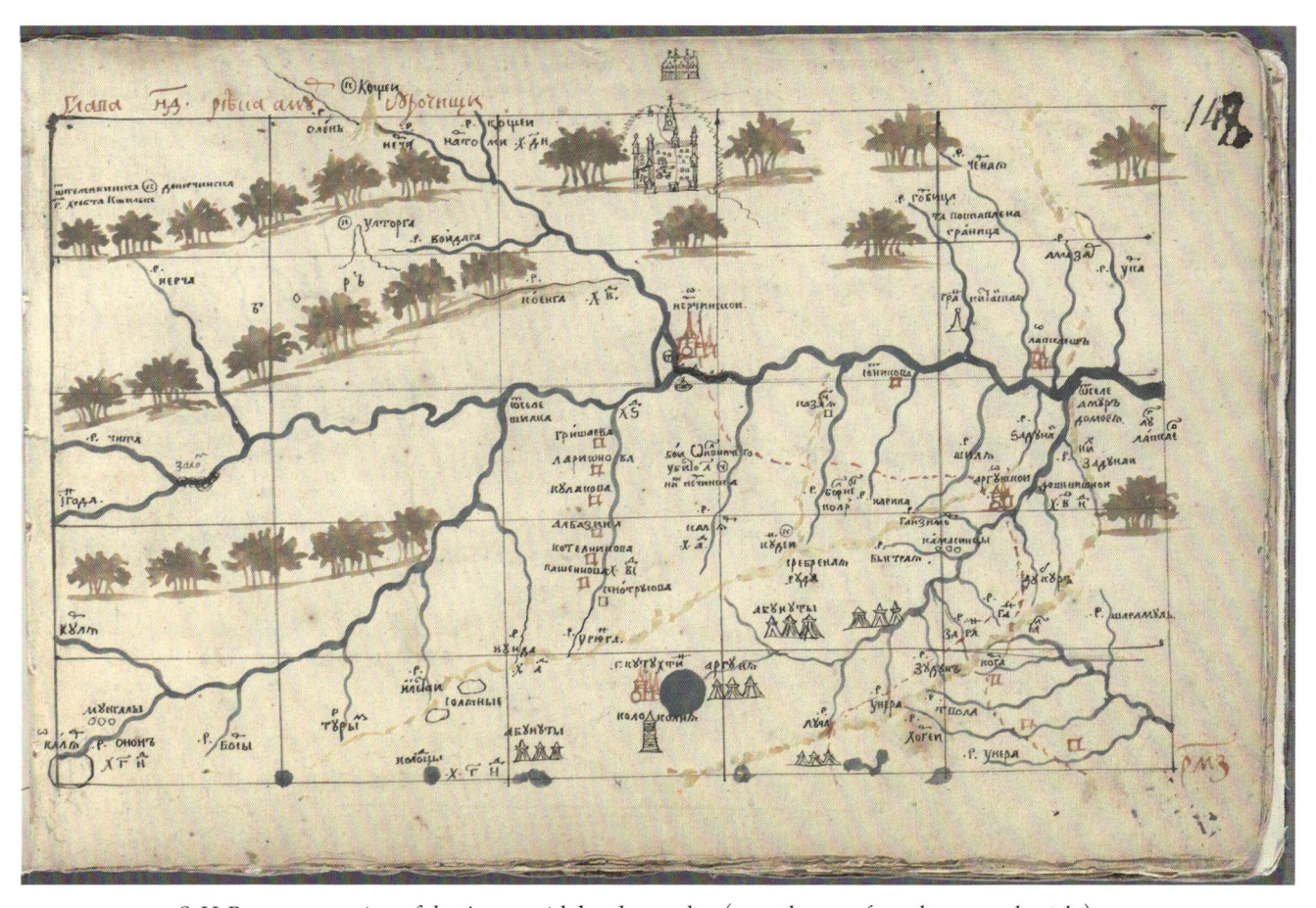

FIGURE 1.3. S. U. Remezov, section of the Amur, with border marker (*second square from the top on the right*), tents, and villages. Nerchinsk appears in red in the top middle, at the branching of the Irtysh and Nercha Rivers. From *Chorographic Sketchbook* (*Khorograficheskaia chertezhnaia kniga*) (1697–1711), 23 × 30 cm. chap. 54, fol. 148. Public domain. Available at https://iiif.lib.harvard.edu/manifests/view/drs:18273155$182i.

D. G. Messerschmidt's expedition, exploring and mapping Siberia from 1719 to 1727. His map, initially published as a stand-alone in 1725, combines information from Remezov's maps and others with the expedition's findings. He emblazoned his map with the title, all in emphatic capital letters, "imperium russicum regnum."[24] A bit later, a map of "Generalis Totius Imperium Russorum," published circa 1750 in an atlas of Johann Homann, splashed across the entire Siberian space the unequivocal label "Imperium Moscoviticum (Siberia)."[25] Remezov and his Russian contemporaries did no such thing.

Throughout the Petrine era, such titulature, with its explicitly political claim making, appeared only on foreign-made maps. Despite the close interactions between foreign and Russian explorer-cartographers, those big-picture political titles with their assertions of uniform territorial sovereignty remained unknown in early eighteenth-century Russian mapping culture.

The first Russian product to wave the flag so conspicuously was Ivan Kirilov's *General Map of the Russian Empire*, published as the grand opening to his atlas of the empire in 1734, during the reign of Empress Anna Ioannovna, Peter's niece. Both the title of the map and unmistakable visual messaging of the cartouche proclaim the empress's sovereign power. Kirilov situates the Russian Empire in the company of other empires and monarchies. Along with the Russian Empire, inscriptions identify Imperii Turcici Pars, Sinarum Imperium, Regnum Tibet, and Japonae Regnum.[26] Among Russian cartographers, Kirilov first embraced a new representation of sovereignty over an unbroken, bounded territory.

If we shift back to the Petrine era, however, before Kirilov's innovation, we have seen that Russian maps registered the sovereign's realm with fuzzy edges and holes in tsarist control rather than by any assertion of a uniform, blanket extension of authority over homogeneous space. What, then, did Russian mapmakers intend to accomplish? Remezov's maps and those of his contemporaries note the *iasak* status of each group to indicate their degree of compliance. An annotation on Remezov's map of the Yakutsk region in the *Chertezhnaia kniga* notes, "Lakes, and along these lakes live many non-*iasak*-paying Tungus who engage in fishing and hunting and trapping sables." The impulse to track the lacework of sovereignty extended to the farthest reaches of the frozen tundra. On the same map he notes, "Here is an impassable peninsula which flows from the top of the Anadyr River to the sea. And the end of it is unknown. On the peninsula live many foreigners (*inozemtsy*, literally, "people of other lands"), non-*iasak*-paying Koriaks and Chukchi."[27]

The allure of fur and the struggles against intrac-table populations surfaced all over the maps. Remezov remarks: "The River Kamchatka, and along it live non-*iasak*-paying Kamchadals. Their clothing is made of dog and sable and fox. And the Cossack officer Dmitrii Potapov was sent from Yakutsk to the Koriaks along that river in 1685/1686."[28] Fedor Beiton, a contemporary of Remezov, notes on his map from 1710/11 of the Enisei River and Kamchatka: "On [this island] the people are all non-iasak paying. Ivan Golygin was there. There are brown foxes and red Siberian foxes."[29] The cryptic nod to Ivan Golygin suggests that an earlier Russian or Cossack effort to claim the region in the name of the sovereign had failed. With this glancing note, the map acknowledges yet another of the gaps that constituted the lattice of sovereignty.

Erika Monahan has found that Remezov used specific symbols to denote groups resistant to tsarist overlordship. For settlements of those same "unpeaceful" Samoyeds we have already encountered, he used a particular indicator—pointed tents, instead of the miniature houses, circles topped with crosses, or tiny forts or towns that he used for other populations. "Throughout the atlas, tents appear where the grip of Russian sovereignty was particularly illusory, or nonexistent. In the entire atlas there are about 220 tents (amid over 2,000 non-Russian population points). These tents appear at the most northern and at the most southern lands shown in the atlas." Monahan posits: "These tents indicate a certain freedom from tsarist control."[30]

Monahan's observation provides important clues to precisely the topic of our volume, territorial imagination. As she says, within Siberia, sovereignty might be tenuous, illusory, or even nonexistent. The challenge to the mapmaker was to figure out how to represent those conditions spatially. Remezov and his contemporaries adopted a strikingly granular approach, a vision of a vast space (about 3,500 miles from the Urals to Kamchatka), a finely parsed human geography of subjugation and refusal. In a seemingly endless landmass, where peninsulas were deemed "without end," deserts and mountains "impassable," and certain people "impossible to subdue," sovereignty, and the maps that recorded and built it, were constructed out of individual relationships. Sovereignty, such as it was, was not a matter of territory. It was established only in the reciprocal acknowledgment by all parties. Where that acknowledgment was not forthcoming, where it could not be negotiated or, more often, violently coerced, the tsar's power stopped short. Hence the astounding commitment, seen in Remezov's maps, to identifying each household in Siberia's vast expanse, whether that of a Russian settler or an Indigenous reindeer herder, and noting its status within a calculus of submission or resistance.

Considering the magnitude of the task, Remezov and his fellow mapmakers managed to incorporate the human population at an astonishingly high resolution, down to the individual households and arable fields of Russian peasant settlers and the *ulus*es, or household units, of nomadic clans. Detailed views show individual houses and settlements. A dramatic example of this detailed perspective is seen in a note inscribed alongside a small picture of two houses on

the map of the Mangazeia region in the *Chertezhnaia kniga*: "The Great Sovereign's agricultural peasants live here, 2 people."[31] Two people! That's pretty fine tuning in a mapping project at Eurasian scale (figure 1.4). To Remezov, they were pinpoints of control in an ambiguous landscape of power, and they embodied the *point*, the goal of his work. The carefully delineated squares of arable fields marked out near the Irtysh River in figure 1.5, from the Chorographic Sketchbook, similarly call out the granularity of these maps. The small rectangles indicated with dotted lines represent plowlands farmed by Russian or Ukrainian settlers. The tents represent the *ulus*es of Indigenous residents. This map is enlivened by details that Remezov, who lived in this region, would have known. Under a peaked yellow rock formation toward the bottom right he has drawn and labeled a collection of bones (unfortunately without explanation). A bit further to the right, he included a line of flowering plants, labeled to tell us they are berries of an unknown sort.

The same attention to individuals and small groups characterizes Remezov's taxonomies and placement of the native Siberians and other non-Russians. In the Chorographic Sketchbook, individual houses and settlements appear along the tributaries of the Amur: squares for Russian villages; tents for non-Russian Arbunuts (see figure 1.3). Hunger for fur would suffice to explain the regime's interest in such individual-level accounting of the *inozemtsy*: to collect tribute, the state needed to know where its tribute payers were, and the vastness of the Siberian taiga and tundra made it all too easy for them to dis-

FIGURE 1.4. S. U. Remezov, Mangazeia region ("Chertezh zemli Turukhanskago goroda") in the *Chertezhnaia kniga* (1701), map 13, fol. 28 (detail). The text near the two little houses in the bottom left says the settlement houses "two people." Public domain. Available at https://kp.rusneb.ru/item/material/atlas-sibiri-semena-remezova.

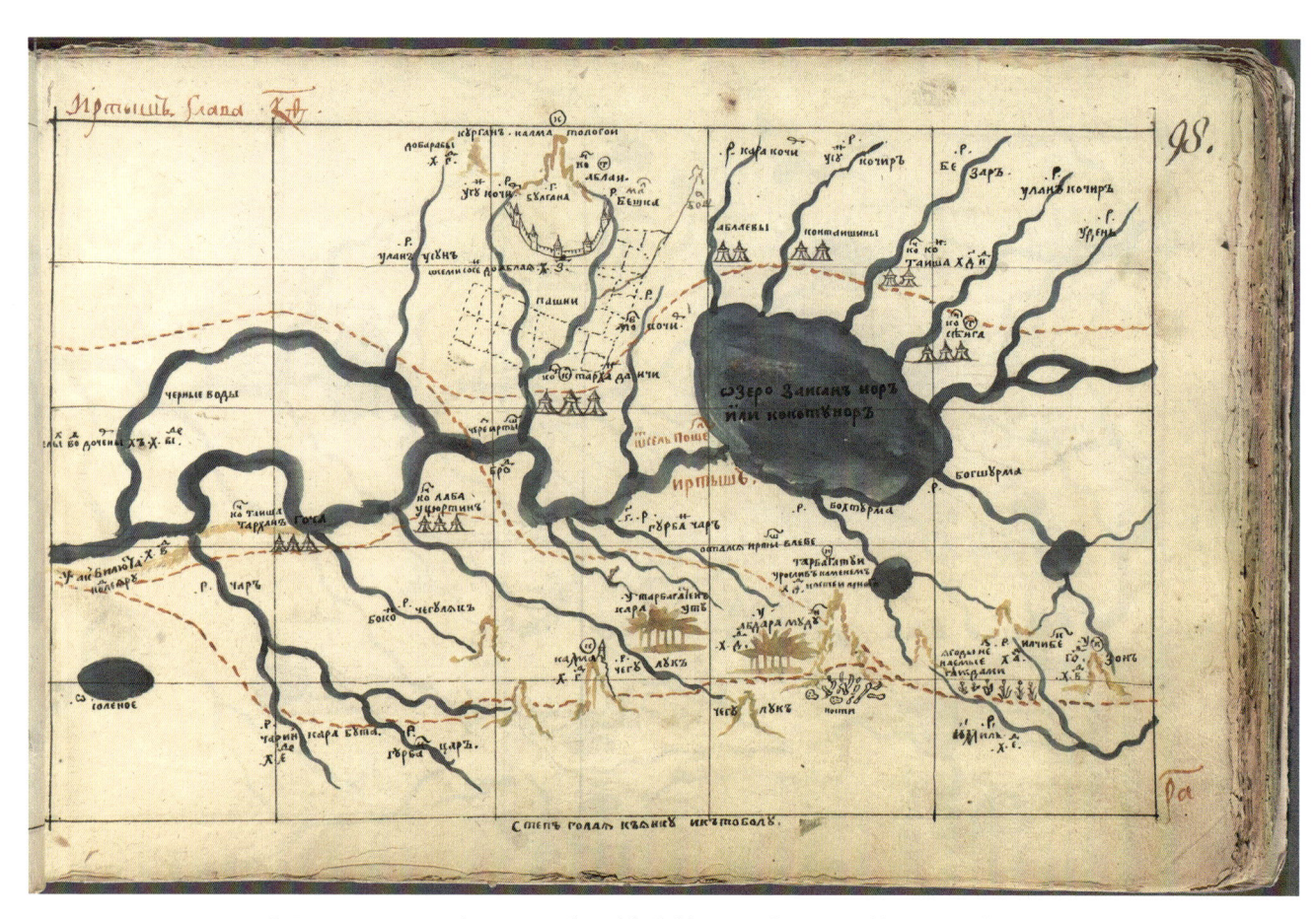

FIGURE 1.5. S. U. Remezov, Irtysh River, with arable fields, tents, bones, and berries. Yellow outcroppings are natural rock formations; the curved wall at top center represents the city of Bulgana. From *Chorographic Sketch Book*, chap. 29, fol. 98. Public domain. Available at https://iiif.lib.harvard.edu/manifests/view/drs:18273155$122i.

appear. To complicate matters further, most of the Siberian peoples were migratory. Maps note that particular groups would "roam (*kochuiut*)" in certain areas. Their settlements are identified with the letter *k*, signifying *kochev'ia*, or nomadic camp.[32] The task of pinning them down in space would seem Sisyphean, counting grains of sand. And surely many of the nominally loyal tribute payers slipped through the cracks. But the maps and the tribute-collection records show a remarkable degree of success at tracking them down from year to year, exacting plush sable pelts through threat or through outright force. The petitions of the tribute payers similarly confirm the efficacy of the tsar's system: they threw themselves on his mercy, describing the miserable state to which the tribute collectors had reduced them.[33]

The state and its cartographic staff were interested in more than locating those who were already firmly "under the tsar's mighty hand." The maps not only provided snapshots of the tribute-paying population but also exposed those who evaded, avoided, and resisted falling into that category, who punctured the overlay of domination. The maps charted the holes in sovereignty with as much individual detail as they tracked those sworn to fealty.

The consequences of such lacunae were spelled out in a note tucked into the Chorographic Sketchbook to explain the scenes depicted on one of the maps: "And above the lake live many Tungusy of various clans. And they pay no iasak to anyone." More of these troublesome people live along nearby rivers: "And along the Ziia river live Reindeer Tungusy and along the Bystraia River live Boat Tungus, and they

live on fish. And the Bystraia joins the Khamun upstream. And to cross from one to the other takes 6 days. And downstream along the Khamun live small tribes of Tungus who pay no iasak to anyone."[34] A sketch of the final stretch of the Amur River before it debouches into the ocean illustrates this wayward population. Tiny red circles signify "free (*vol'nye*) Tungus of various clans (*roznye rody*)." Their "free" status is not listed as a positive characteristic. Rather it conveys their refusal to bow to the tsar's might and their refusal to fork over tribute in fur.[35]

The striking lack of representation of homogeneous, horizontal, bounded territoriality on the maps reminds us that sovereignty was understood in fundamentally different ways in the Russian Empire of the early eighteenth century. Differing as much from ideas developed in early modern Europe as from the post–World War II system of sovereign states, in the Siberia of Peter the Great, sovereignty was about minimal subjugation and nominal loyalty, and most of all, it was about piles of soft, warm sable skins.

Throughout this discussion, I have used a variety of metaphors to try to capture the kind of indeterminate sovereignty expressed in Petrine era sources. I have invoked images of lacework, of holes or lacunae, ambiguity, and tenuousness. And I think these point us in the right direction. At the start of the eighteenth century, the empire was not imagined as an integral, homogeneous, bounded territorial entity. It did not exercise authority across a uniformly subjugated space. It could not command the loyalty, or even the nominal compliance, of a homogeneous population.

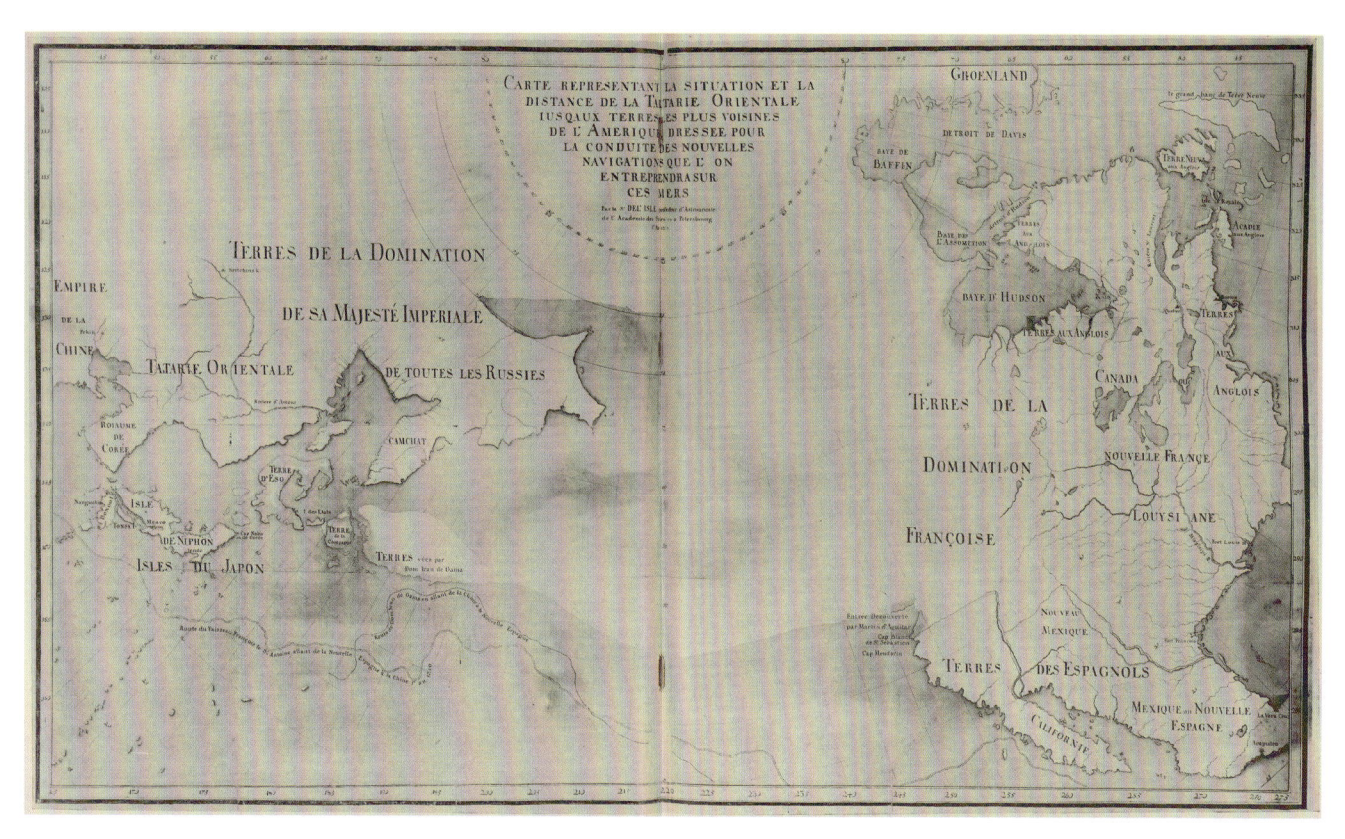

FIGURE 1.6. Joseph-Nicolas Delisle, *Carte representant la situation et la distance de la Tartarie Orientale jusqaux terres les plus voisines de l'Amerique, dressée pour la conduite des nouvelles navigations que l'on entreprendra sur ces mers*, composed in 1733. Note the "Terres de la Domination de sa Majesté Imperiale de toutes les Russies." Public domain.

Nor did it particularly aspire to assert blanket authority over contiguous, bounded space. The maps illustrate the patchiness of tsarist control even within the nominal, invisible borders of the empire.

The first map to express a vision of the Russian Empire as a continuous territorial expanse was created by yet another foreigner, Joseph-Nicolas Delisle. An accomplished cartographer from a distinguished mapmaking family in France, Delisle accepted an invitation from the tsar to relocate to the Russian Academy of Sciences in St. Petersburg and to work on the grand project of mapping the empire. His work was well received, and many of his maps are well known, having made their way into circulation in the west. Less familiar is his hand-drawn *Carte representant la*

situation et la distance de la Tartarie Orientale jusqaux terres les plus voisines de l'Amerique, dressée pour la conduite des nouvelles navigations que l'on entreprendra sur ces mers, composed in 1733 between the first and second Bering expeditions (figure 1.6).[36]

Visually, the map is phantasmagorical, disorienting. Land and sea blur into each other; continents appear out of the mist. Yet the map moves toward unambiguous statements about political dominion. It portrays sovereign claims in an almost Westphalian mode, as uniform, uninterrupted dominion over contiguous space. Like the other European cartographers whose work we have examined, Delisle proclaimed Russian sovereignty in Siberia with his title: *Terres de la Domination de sa Majesté Im-*

periale de toutes les Russies. What is new here is that the map is not only unequivocal labeling, but it is equally definitive in its visual message. Russia appears as one and indivisible, under the rule of "her Majesty," the empress Anna. Her "lands" stretch in unbroken continuum from the westernmost edge of the map (somewhere unspecified in Siberia) to the Pacific and from the Arctic to China. Like some of his predecessors, Delisle places the "Lands of the Domination of Her Majesty" among other empires and kingdoms. However, the list of relevant powers reveals a very different world order, radically changed in just a few years. Instead of the regional players—Korea, Giliatskaia, and the Tsardom of China—the cast of characters is largely European. Across an indistinct stretch of the Pacific Ocean, Russia's empire is flanked in North America by "Lands of the French Domination," "Lands of the Spanish," and, crammed in the upper-right corner, "Lands of the English." Still in the mix, closer to home, are the "Empire of China," the Kingdom of Korea, and the island of Japan.

Like earlier European mapmakers, Homann, de Fer, Strahlenberg, and Delisle used words to estab-lish the monarch's possession of the territory. But Delisle went further, effacing all divisions, subunits, and unincorporated spaces that might interrupt the empress's domain. His map makes an unambiguous visual statement and a claim for absolute geographic unity under a majestic sovereign. Russia, as filtered through Delisle's vision, had entered a modern world of contending imperial pretensions in a European mode. The hardening of notions of territorial sovereignty took shape in a world of empires. European cartographers shaped visions of Russia as imperial power, while border stones, restriction of cross-border movement, and diplomatic treaties solidified boundaries in the Far East. With Russia riding roughshod over Ukrainian sovereignty today, the idea of a solid, internationally recognized, mappable boundary line that would secure the territory of an autonomous, inviolate state starts to take on newly positive, even aspirational meaning. The task of refining our understanding of historical relations between people, sovereignty, and territorial imaginaries becomes all the more important.

2

Lhamsuren Munkh-Erdene

From People to Territory
(The Chinggisid) Sovereignty Transformed?

Mainstream political theory is based on an evolutionary narrative. It posits that political authority underwent a fundamental transformation in recent centuries, shifting its focus from people to land. This supposed people-to-territory transformation is seen as signaling the advent of the modern international system: whereas premodern political authority was exercised over people and built on personal ties, modern political authority by contrast is exercised over territory. People fall under its sway, to be sure, but "by virtue of living on [a] piece of land."[1] This suggests a sharp divide between *"territorial sovereignty"* and *"tribe-*sovereignty"—the Westphalian and the pre-Westphalian.[2] The divide is not just between different forms of political authority; it is the "Great Divide" between the state of nature and the condition of the state.[3] The divide is at once environmental, economic, evolutionary, and developmental, making it the greatest of divides in human history. Nomads, in this account, have no sovereignty, for they occupy land "only as a periodic route of migration rather than as permanent and exclusive claims."[4] If the epitome of nomads is the Mongols, the most

often cited authority on nomadic attitudes toward land is Owen Lattimore, whose work epitomizes this approach.

The present essay critiques this intellectual legacy in two ways. First, I trace the convoluted history of the people-to-territory idea from the seventeenth to the twentieth century. As we will see, the Great Divide is a product of evolutionary materialist thinking. Henry Maine's tribe sovereignty was a construct of colonial ideology; when Lattimore applied the same scheme to Mongolia, it carried a similar colonial agenda. Yet in Inner Asia as elsewhere, the historical record shows that political authority has always been exercised over people and has always had territorial limits. This record is explored in the second part of the chapter, where I delve into the rich chronicles of the Mongols themselves. The actual transformations we find there as in Europe are more subtle: from divine majesty to popular sovereignty, via princely sovereignty; from traditional authority to legal authority; and from patrimonial-feudal government to the kind of republican government that created the *Rechtsstaat* based on rule of law.

The Great Transformation

The idea that sovereignty underwent a fundamental shift from a premodern focus on people to a modern focus on territory is the essential Westphalian story—a story that came to be projected worldwide during the course of European colonialism. Westphalian sovereignty was a princely prerogative for autonomous self-rule within an imperial framework.[5] Recognizing multiple sovereigns, the Westphalian system created a constitution that made legal equality and noninterference among the treaty's signatories its constitutive norms; dismantling *imperium*, it created sovereignty as a political right and authority bound by these norms. Since territory—that is, bounded land under jurisdiction—embodied these norms, Westphalian sovereignty became known as territorial sovereignty; lesser principalities acquired a legal recognition for autonomous existence within the system.

As a hard-won right, Westphalian sovereignty was a prerogative originally limited to the treaty signatories; no signatory was to recognize the sovereignty of a nonsignatory, for power politics was a zero-sum game. This rendered sovereignty an exclusive monopoly of a handful of European rulers, turning the rest of the world into *territorium nullius*—land to be claimed by these sovereigns upon discovery or conquest.[6] On the one hand, making the signatories sovereign in their respective jurisdictions allowed these princes to consolidate power at home through eliminating patrimonial-feudal authorities within their territories. This pro-cess eventually resulted in nation-state territories with hierarchically integrated uniform administrative divisions. On the other hand, the "doctrine of superiority of agriculture" entitled the same European powers to justify extending their sovereignty over the rest of the world.[7] While some early writers on international law had regarded "backward races as possessing a title to the sovereignty over the territory they inhabit which [was] good as against more highly civilized peoples," most recognized no such Indigenous claims.[8] Instead, the lands of non-Westphalian rulers were conceived as commons to be claimed by Westphalian sovereigns.

This justification for empire was clearly articulated by the political theorist John Locke. In order to make Native American territories commons to be appropriated by those who till, plant, improve, and cultivate land, Locke reduced the "*Kings* of the *Indians* in *America*" to "little more than *Generals of their Armies*."[9] Centuries later, pioneering anthropologists like Henry Maine and Lewis Morgan went further. Claiming that "the king of a whole tribe was king of his people, not of his people's lands," and further that "territorial titles were not known" in the Americas, Maine argued that "*tribe*-sovereignty" had "no claim of right upon the fact of territorial possession, and indeed attached no importance to it whatever."[10] In contrast to tribe sovereignty, modern territorial sovereignty was defined as "the proprietorship of a limited portion of the earth's surface," where "sovereigns *inter se* are to be deemed not paramount, but absolute, owners of the state's territory."[11] Here we see the hardening of the distinction between two forms of political authority. For Maine, "Kings of the French" were fundamentally different from "Kings of France."[12] What separated the two was the proprietary right to land—and it was this proprietary right above all that was coming to define sovereignty. In effect, sovereignty was being recast as a matter of proprietorship rather than rulership. Only those who till, plant, and cultivate land were seen by Maine as exercising true sovereignty.

Adding a kinship dimension, Morgan widened and deepened the divide. Dividing all forms of government into either "society" ("founded upon persons, and upon relations purely personal") or "state" ("founded upon territory and upon property"), and postulating that the latter had evolved out of the former, Morgan transformed Maine's two forms of sovereignty into distinct stages of political organization: ancient and modern.[13] Ancient society in this view was fundamentally kinship society, "founded upon the kinship of the gentes" and "formed by the segmentation" of lineage descent.[14] Morgan thus posited a "transition from gentile into political society."[15] The evolution of society was coming to be narrated as "a development from kinship to territory as basis of association."[16]

Morgan took an additional step as well, singling out agriculture as the practice that had propelled humanity to "the civilized state."[17] If economy determined human social and political organization, it was agriculture that separated civilization from barbarism.[18] In Friedrich Engels's famous formulation, social evolution was "the progress in the production of the means of sustenance"—in a word, "control over

food production."[19] For these influential thinkers, food production and population pressure had fueled human history, driving progress, evolution, and civilization. As slavery and the state were the creations of wealthy and populous societies, so civilization and sovereignty were the fruits of agriculture. This turned the divide between tribal society and territorial state into an evolutionary one with "ascending" and "descending" stages.[20] Such an account suggested, moreover, that the existing order was inevitable.

Since the publication of Meyer Fortes and E. E. Evans-Pritchard's influential *African Political Systems* (*APS*) in 1940, ethnography has repeatedly proved Maine and Morgan wrong. Arguing that territorial frameworks are universal, and that the real distinction is whether a society is governed by administrative pyramid or kinship segmentation, *APS* rejected the long-standing orthodoxy. Both the state and kinship organizations turn out to be territorial; what differentiates them is the presence or absence of "the organized exercise of coercive authority."[21] In the state, "the administrative unit was a territorial unit"; in the kinship systems, "the territorial units . . . corresponded to the range of a particular set of lineage ties and their bonds of direct co-operation."[22] What the segmentary system lacked was not territory but an administrative system; the transition became one from kinship-to-bureaucracy.[23]

APS also revealed that there was no correlation between subsistence systems and political organization. On the one hand, many nomadic pastoralists had created states; on the other hand, some sedentary agricultural groups had no state. Neither did population size correlate with political organization: "A stateless political unit need [not] be very small . . . nor a political unit with state organization need be very large."[24] Finally, no consistent correlation was found between population density and political organization. While groups with low population density like the Zulu, the Ngwato, and the Bemba built states, some societies with much higher density like the Nuer, the Tallensi, and the Logoli retained a segmentary lineage system.[25] In short, from Yąnomamö villages to New Guinea highland hamlets, all human societies turned out to be territorial, confounding Maine's notion of tribe sovereignty.[26] Even other primates proved to be territorial.[27] Evolutionary anthropology thus had to abandon the kinship-to-territory transition, accepting territoriality "as the basis of association . . . as old as consanguinity."[28]

Yet even as evolutionary anthropologists abandoned the kinship-to-territory transition, they clung to the idea of the kinship-to-bureaucracy one. Administrative divisions and bureaucracy emerged as keys to defining the state, leading many scholars into the new error of thinking that only where there is bureaucracy could there be a state. In fact, bureaucracies cannot create themselves. Rulership and appointment are mutually exclusive principles, as Max Weber knew. For Weber, "the purest type of exercise of legal authority is that which employs a bureaucratic administrative staff"—officials who are "*appointed, not elected*."[29] Yet "the supreme chief of the organization occupies his position of dominance (*Herrenstellung*) by virtue of appropriation, of election, or of having been designated for the succession."[30] The

sovereign, in other words, "is not selected according to the same norms as the officials in the hierarchy below him. Exactly the pure type of bureaucracy, a hierarchy of appointed officials, requires an *authority* (*Instanz*) which has not been appointed in the same fashion as the other officials."[31] The state, in short, is not defined by bureaucracy after all.

Inner Asian Frontiers of China

The Mongols have long been viewed as one of the latest societies to make the transition to territorial rule, supposedly during the era of Daiching (Qing or Manchu) rule (1644–1911).[32] According to Owen Lattimore, the Manchus accomplished a social revolution in Mongolia by establishing *khoshuu* (principalities) with definitive territorial boundaries.[33] "The tribal following of a ruling prince—the tribal unit, that is, which the Manchus, by defining territorial boundaries, converted into something resembling a petty state, or principality—is the *hoshio*."[34] Under the Qing, Lattimore argued, the Mongol prince for the first time "became a petty sovereign, at once restrained and supported by an overlord."[35]

If this were true, the Mongol transition would paradoxically have run in the opposite direction to the Westphalian transition. Whereas petty lords in Central Europe, repealing the *imperium* of the Holy Roman Emperor, claimed sovereignty for themselves at Westphalia, in Mongolia (according to Lattimore) the Daiching *imperium* itself created petty states and sovereigns within its empire, vis-à-vis the emperor. One might construe this as a Manchu civilizing mission, but Lattimore conceived it as a Chinese civilizing mission, positing that "the economic, social, and cultural power of the Chinese themselves, as well as their political energy, radiated beyond their own containing Great Wall into the steppe."[36] This causal logic is convoluted and counterintuitive in several ways, not least in that it casts the conquered and bureaucratically administered Chinese subjects as the original creators of aristocratic principalities in Mongolia!

No transition could have been more dramatic than this: the power and authority of the Mongol khan or the Great Khan of the Mongol Empire that had once epitomized sovereignty for European thinkers—"the most high, absolute, and perpetual power over the citizens and subjects in a Commonweale" for the original theoretician of sovereignty[37]—were recast by evolutionary anthropologists as the diametrical opposite of sovereignty, to the point that the Mongolian political organization could be described as acephalous and lacking "nerve-centers."[38] By thoroughly tribalizing Eurasian nomads in this way, colonial anthropology managed to reconfigure the Eurasian steppe as the "Inner Asian frontiers of China": a classic *territorium nullius*, a "barbarian 'outer darkness'" akin to James Scott's Zomia.[39] The difference was that, for Mongolia, the civilizing force was said to come from an Asian empire rather than a European one.

In the nineteenth century, Henry Maine had attributed his tribe sovereignty's disregard of territory to the traditions of the "nomad horde, [which] merely encamped for the time upon the soil which afforded them sustenance."[40] For Lattimore, this same "ancient pure nomadic instinct" and "primitive concept of sovereignty . . . in which land has no place at all" could still be found among the Mongols in the 1930s.[41] "The basis of Mongol land tenure," Lattimore claimed, "is that all land belongs to all the tribe, even the prince having no prescriptive right."[42] Yet he also argued that "the basic feeling of the Mongols is that the land belongs to the whole tribe. Neither individuals nor the chief may establish a prescriptive claim to personal ownership of any part of the land."[43] Thus, there *was* ownership of land among the Mongols. The "tribe," or the body politic, owned the land and exercised sovereignty over the territory just as in other polities, although it prohibited any private ownership of land, including by the prince. Mongolia's sovereignty was absolute and indivisible, and it was the body politic that had the absolute and indivisible right to the land.

Lattimore mischaracterized not only the Mongols of the 1930s but their predecessors as well. His historical claims are refuted by many early observers, from Sima Qian and *The Secret History of the Mongols* (*SHM*) to William of Rubruck, John of Plano Carpini, and Boris Vladimirtsov.[44] While Lattimore sometimes quotes their texts, all these authorities squarely contradict his claims; for example, while Lattimore mentions William of Rubruck and John of Plano Carpini, he is glaringly silent on the weighty remarks that both William of Rubruck and John of Plano Carpini left on territorial arrangements of Mongolian decimal divisions.[45] Likewise, starting from 1947, Lattimore often cited *SHM*, yet he never engaged with its account of decimal organization or the allocation of territories to thousands (*minqan*). Lattimore also never acknowledged *SHM*'s numerous passages that discuss the key term *nuntuq*.

This term deserves our attention. *SHM* employs two words, *qajar* and *nuntuq*, that have similar but distinct valences. The *qajar* (land, earth) does not necessarily have an owner.[46] In contrast, *nuntuq* (territory, country) is owned, protected, granted, and expanded. According to *SHM*, "sons protect *nuntuq*, [while] daughters get selected by their beauty."[47] While all the named communities (or polities) have their own *nuntuq*, *nuntuq* can be subdivided, allocated, granted, and expanded.[48] It is the ruler—the khan, "the sovereign of the nation" (*ulus-un ejen*)—who "allocates *nuntuq* to his subjects."[49] The Chinggisid polity as a whole formed a *nuntuq* too; Chinggis Khan is said to have decided to "expand the *nuntuq*" (*nuntuq a'utkin*) to make Jochi and Chaghadai "rule aliens."[50]

I have taken the trouble to trace this scholarly history because of the lasting impact of Lattimore's work. As Mark Elliott recently noted, "Lattimore is one of the few Asianists of the twentieth century whose influence continues to be felt widely across all fields of history, in large measure because he pioneered an approach to the study of the frontier."[51] According to this Harvard historian, Lattimore "stands as the intellectual godfather of all historians of China" and its frontiers; "for the beginning student of the Chinese frontier, Lattimore is often still the first author we read"—one whose "writings continue to provide an intellectual context in which to think about the frontier in general, both across a broad temporal sweep and in comparative terms."[52]

What is not noted by Elliott is that Lattimore's oeuvre, especially his *Inner Asian Frontiers of China*, was driven by an overt objective of "integrating China and its Inner Asian hinterland."[53] The aim of Lattimore's work was to cast "Manchuria, Mongolia, Chinese Turkistan, and Tibet" as Inner Asian frontiers of China—a Chinese hinterland—and to demonstrate "historically and geographically" that these territories constituted an integral part of China.[54] To claim Inner Asia as an integral part of China, Lattimore appropriated the dominion of the Daiching Empire as a Chinese historical imperial dominion, with China proper as its center. Making Manchuria, Mongolia, Tibet, and what is now Xinjiang compose the Inner Asian frontiers of this dominion required making these areas *territorium nullius*, which in turn required conceptualizing them as tribal—that is, without sovereignty or state.

As the remainder of this chapter will show, however, Lattimore's petty states and petty sovereigns were not the Daiching emperor's creation. Rather, they were the work of the *taiji* government: an Indigenous steppe regime that created both the Great Code order and the Daiching Empire itself.

The Great Code Order

The Daiching Empire came into existence from a world very similar to the one that led to Westphalia.[55] In 1640, twenty-eight Inner Asian princes—whose territories stretched from the Volga in the west to the

Hülün Buir in the east, and from Köke Nuur in the south to Krasnoyarsk to the north—met in a Great Assembly (*yeke čiɣulɣan*) and adopted a legal document called the Great Code (*yeke čaɣaja*) with 120 provisions.[56] The same Great Assembly created the so-called Döchin and Dörben: a confederation of principalities, bound by a common legal system under the Great Code (which they mutually enforced among themselves) but otherwise autonomous.[57]

The Great Code not only prohibited any attack against the "greater and lesser states"; it also explicitly banned the "killing and raiding of different communities of lamas and their teachings."[58] As such, it did more than recognize the independence and integrity of principalities; it also accommodated all the Buddhist denominations within the Döchin and Dörben. I have argued elsewhere that the Great Code created a political order akin to that of Westphalia.[59] If anything, the Great Code went further. Whereas even the modern UN system—by creating a security council where all meaningful decisions require the concurring votes of the permanent members—effectively established a *de jure* hierarchic order and "an oligarchy of five men," the 1640 Great Code sanctioned neither hierarchy nor oligarchy.[60] Not only did it criminalize war, it called for complete destruction of any aggressor by collective force.

In contrast to today's UN system, the Great Code discriminated overtly between greater and lesser principalities. Yet it did so precisely to hold greater princes to higher standards. While giving equal protection to all, it imposed heavy punishment and penalties on aggressors, with heavier punishment for more powerful aggressors. The enforcement was collective, with weightier penalties for greater principalities whose rulers failed to take part in the enforcement. If a greater state were to be invaded and conquered, the aggressor state was to be divided in two, giving half to the injured party as a restitution. Since conquering a greater state requires greater force, the aggressor in this case was likely to be a major power. By contrast, if a peripheral lesser state were to be invaded and conquered, all its losses in humans and animals were to be compensated by the aggressor. Additionally, the aggressor was to be penalized by 100 suits of armor, 100 camels, and 1,000 horses. In the enforcement of the code, the penalty was 100 suits of armor, 100 camels, and 1,000 horses for greater princes who failed to take part, compared to 10 suits of armor, 10 camels, and 100 horses for lesser princes. "Where there is great power there is great responsibility, where there is less power there is less responsibility" is the principle behind the Great Code; the whole logic is to tame the greater powers.[61]

These provisions made the Great Code order highly favorable to small states. Since the princely assembly made all the major decisions—including recognition of new principalities—the mandatory participation of all the princes guaranteed the rule of the lesser rulers.[62] Notably, the regime created by the Great Code outlasted the confederation that authored it, remaining intact throughout the era of Daiching rule (1644–1911). In 1661, when Lubsang invaded the Zasagtu Khanate—killing the khan and seizing his subjects—the response was decisive and swift: in keeping with the Great Code, the Döchin and Dörben princes jointly subjugated Lubsang, confiscated his subjects, and restored the Zasagtu Khanate.[63]

The Great Code order was an outcome of a process akin to that which took place in seventeenth-century Europe. The disintegration of the Mongol Empire entailed fragmentation by patrimonial divisions, resulting in federalization similar to that of the Holy Roman Empire. The division of the Mongol Empire into princely shares (*qubi*) under the Dayan Khanids transformed the empire into a commonwealth of princely appanages.[64] The process fundamentally transformed the Chinggisid concept of the sovereign (*ejen*), reducing the khan from ruler-and-owner to ruler only. Chinggis Khan, like Dayan Khan after him, had ruled *and* owned the realm.[65] But Dayan Khan's installment of his sons over the divisions of the realm, and his establishment of the principle of primogeniture, transformed the khan's sovereignty. This was the triumph of what a nineteenth-century Mongolian author called the *taiji* government: hereditary aristocratic rulership of princely appanage holders. The Dayan Khanid *taijis* emerged as sovereigns of their respective shares. As a result, though the khan remained "the lawful great khan," he was reduced to mere "sovereign of the throne" or "the sovereign of the title."[66]

The *taiji* government fundamentally transformed the khan as sovereign; the khan could henceforth command but not appropriate princely appanages. All princes were sovereign (*ejen*) proprietors of their shares. Since Dayan Khan installed his sons over the existing political-administrative divisions, they were, first and foremost, rulers of their respective divisions. Over time, however, princes effectively appropriated their administrative districts as their personal appanages and began to subdivide them to their offspring. With an evermore bourgeoning princely population and evermore division by share, the *khoshuu* was gradually established as the effective unit of government.[67]

This process, set in motion to curtail royal power and to protect the interests of princely appanages, resulted in an aristocratic parliamentary government, eventually reducing the khan to the first among equals. Greatly frustrated with the situation, Ligdan Khan (1604–1634) made an attempt to restore the power of the khan. His undertaking backfired, however, resulting in the demise of the Mongol Empire. From its ashes arose both the Daiching Ulus and the Döchin and Dörben confederation.

When Ligdan abolished the *taiji* government, the Dayan Khanid princes revolted and allied with the Manchu lords. All their treaties are treaties of alliance and are built on the principles of equality, mutual respect of sovereignty, and the territorial integrity of the parties.[68] The concept "Whose territory, his jurisdiction" became the basis of their relationship. According to the code adopted on May 31, 1631, if a Manchu were to commit a crime on Mongolian territory, he would be subject to Mongolian laws, and if a Mongol were to commit a crime on Manchu territory, he would be subject to Manchu laws. If a crime were committed on the border of the two states, those involved would be subject to their respective laws.[69] It was these treaties and these principles that created the Daiching Empire as a polity made up of a multitude of lords.

The *taiji* government also fundamentally shaped the Daiching political order, including the Manchu polity. Nurhachi, the founder of the Aisingioro house, apportioned his realm into princely appanages called *hošo*s (Mongolian *khoshuu*).[70] In succeeding Nurhachi in 1626, Hong Taiji made a solemn pledge to his brothers not to seize their appanages.[71] Later, in 1643, Hong Taiji reported that Amin had demanded that his domain should be like a Mongolian principality or "separate dominion."[72] Hong Taiji also said he would be left "without a country" to rule if he let the Manchu princes keep their autonomy; all the Manchu princely *hošo*s were to become external principalities like the external Mongolian principalities.[73]

The division by inheritance and the *taiji* sovereignty fragmented Mongolia; by the end of the Daiching era, there were nearly two hundred principalities. Mongolian princes were dividing their appanages among their sons and brothers, resulting in the clearly bounded units depicted in figures 2.1–2.4. It was these *khoshuu*s that Lattimore deemed as petty states and their rulers as petty sovereigns.[74]

When the Daiching government adopted its New Policies in the early twentieth century, the Mongolian princes declared Mongolia's independence, faulting the Daiching for revoking the established constitutional regime of "original laws."[75] According to the original laws, "Mongolian khans, kings, and ruling princes, being sovereign of their own subjects and owners of their own country and its produce,

have been living very peacefully."[76] In the view of the Mongolian princes, their sovereignty consisted of rulership of their own subjects and ownership of their own territory. Territorial sovereignty of the kind shown in the maps reproduced here was not a Chinese import; it was homegrown in Mongolia.

The Chinggisid Sovereignty

This account is supported by *SHM*, which begins its political history of the Mongols with the phrase "Qabul Khan was ruling Qamuq Mongolia."[77] Three generations later, "Heaven and Earth agreed that Temüjin (a.k.a. Chinggis) shall become the *ejen* of the *ulus*."[78] In making Temüjin the khan, the other princes "pledged their words" and "swore the oath":

> In the days of war,
> If we disobey your command,
> Deprive us of all our goods and belongings, and
> Our noble wives, and cast
> Our black heads on the ground!
> In the days of peace,
> If we violate your counsel,
> Cut us off from our retainers and possessions, and
> Our wives, and cast us
> Out into the wilderness![79]

This oath gave Temüjin the legitimate authority to execute the oath takers for their alleged breach of their oaths.[80] When Chinggis Khan named Ögödei as the next khan, he reminded both Jochi and Chaghadai that Ögödei commanded the same legitimate

FIGURE 2.1. *The Khoshuu of Dugartsembel, the Jorigtu Jasag, the Setsen Khan Aimag,* December 10, 1843 (Daoguang 23rd year). This colorful image depicts one of the smallest principalities in Daiching-era Mongolia. This particular district was legally recognized by the empire in 1755 as the domain of Prince Tserendoyod and his heirs, after the prince donated 1,100 horses, 100 cattle, and 1,000 sheep to the Daiching army during its war against the Zhungar Khanate. Drawn nearly a century later, the map shows the principality as a fertile and well-watered landscape, but also as a clearly bounded polity. As is true of all the *khoshuu* maps reproduced in this collection, the territorial border has been prominently painted, consisting of short, straight lines connecting a continuous series of dots or triangles around the edge of the painted terrain. The dots and triangles represent stone piles, or cairns (*oboo*), erected at strategic points (often at passes or ridges) to help officials on the ground locate the boundary between administrative districts of the Mongolian realm. Each cairn had a name; those toponyms, too, are noted on the map. Courtesy of the National Central Archive of Mongolia.

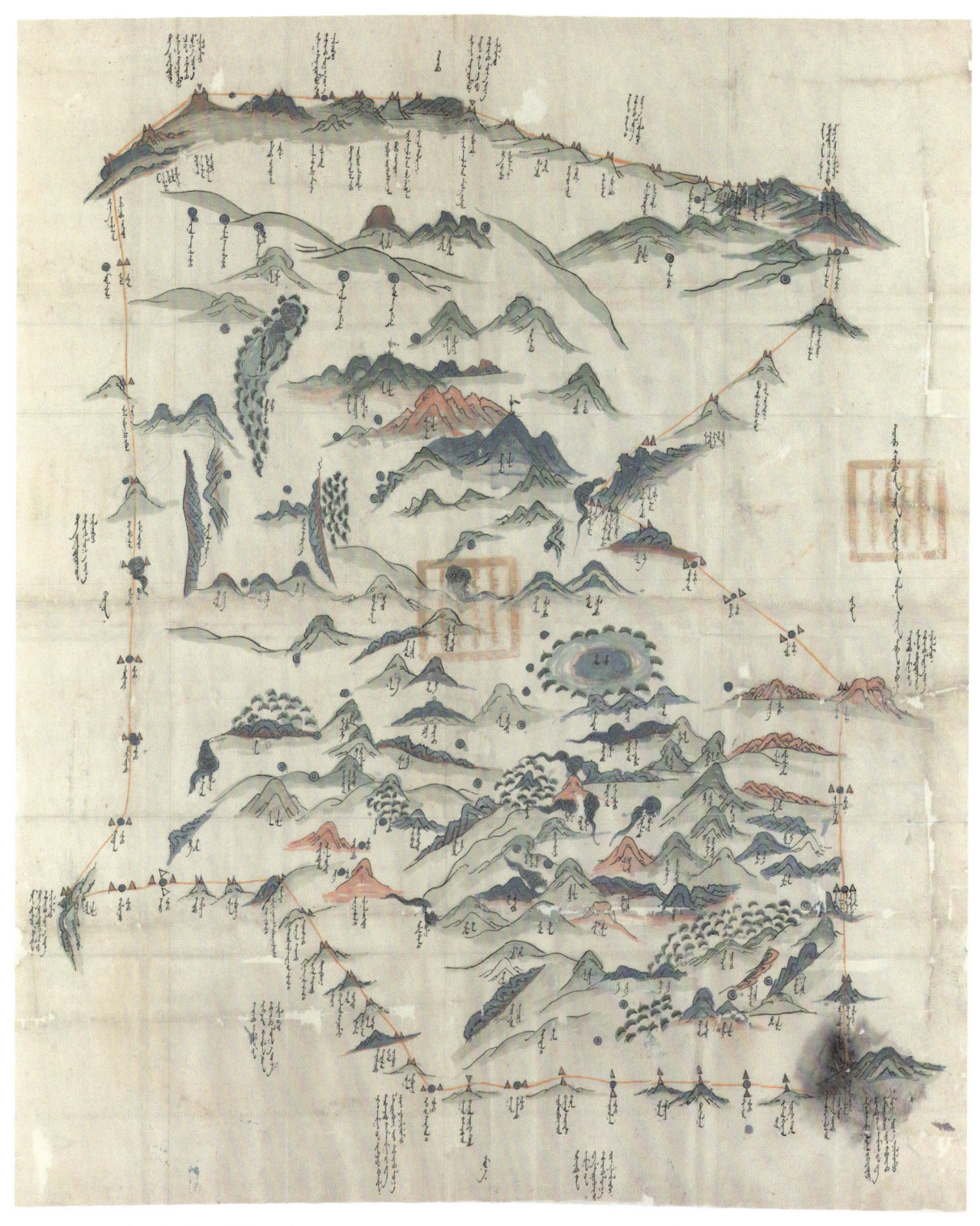

FIGURE 2.2. *The Khoshuu of the Bishreltü Jasag, the Setsen Khan Aimag*, July 15, 1843 (Daoguang 23rd year). This principality belonged originally to a prominent descendant of Chinggis Khan: the Grand Duke Dari, a grandson of Sholoi Mahasammata Setsen Khan (1577–1652). Its last prince was made a king, and the kingdom survived until 1923. This splendid map, produced during the reign of the eighth prince in 1843, marks the border of the principality with sixty-five named cairns. Courtesy of the National Central Archive of Mongolia.

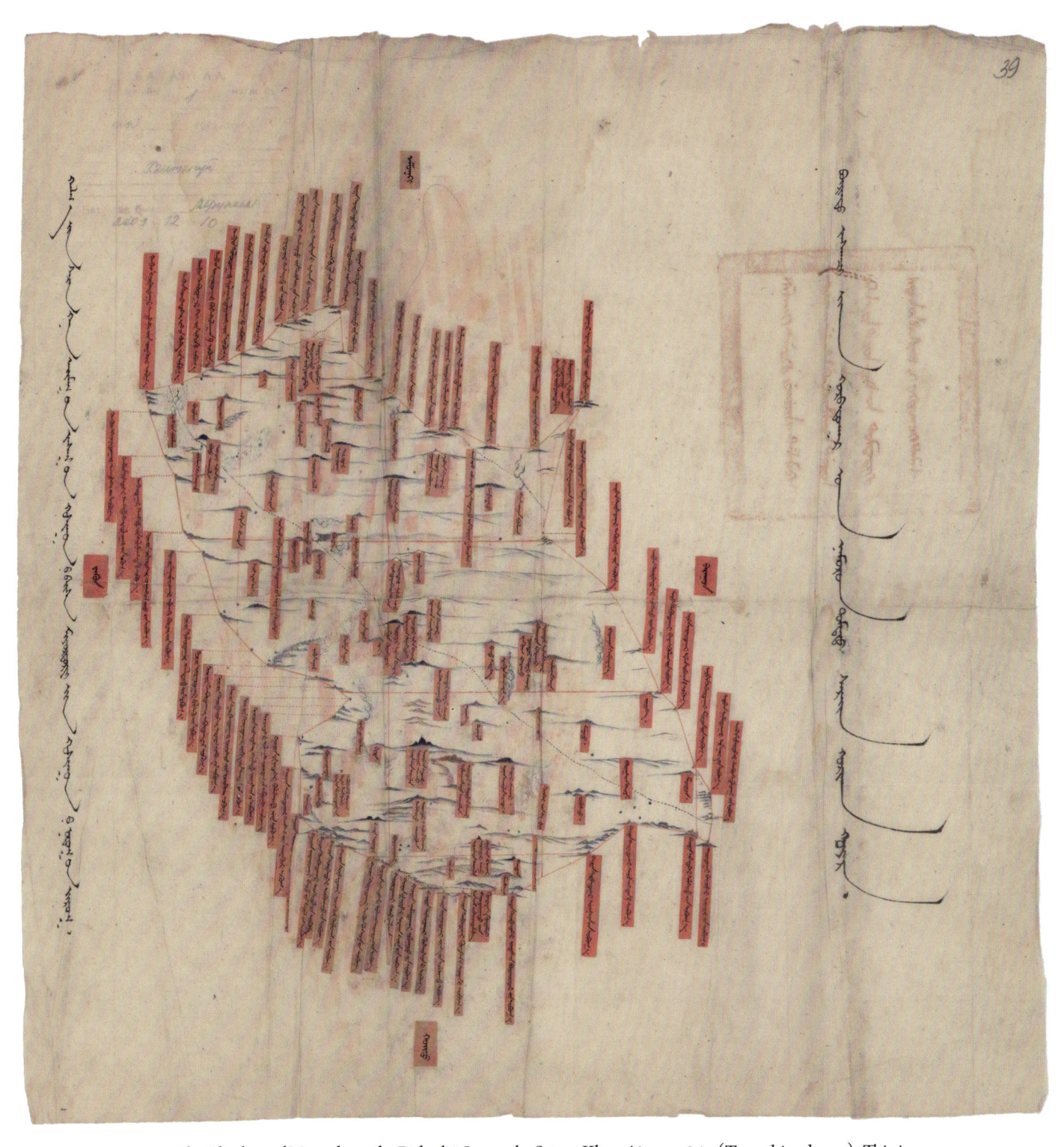

FIGURE 2.3. *The Khoshuu of Miyürdorji, the Bishreltü Jasag, the Setsen Khan Aimag, 1864* (Tongzhi 3rd year). This image, made twenty years later than figure 2.2, depicts the same domain. The content has been simplified, the figure rotated by 180 degrees (putting south at the top), and red textboxes added around its edges. Surrounding the principality on all sides, these banners make the names of its sixty-five boundary markers (*oboo*) the most eye-catching feature of the map. Courtesy of the National Central Archive of Mongolia.

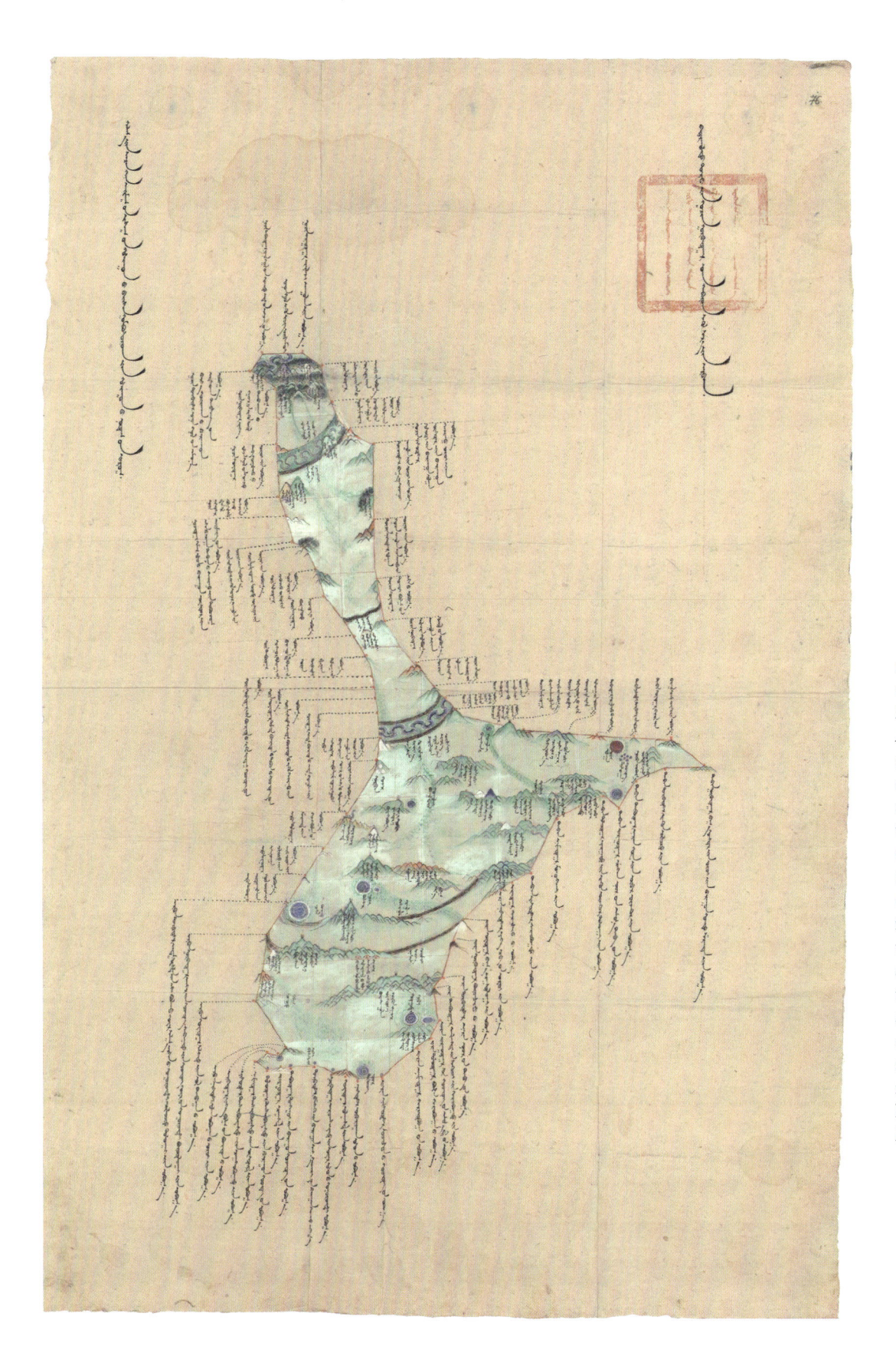

FIGURE 2.4. *The Khoshuu of Gombosüren, the Erdeni Dalai Jasag, the Setsen Khan Aimag,* 1907 (Guangxu 33rd year). Established sometime in the mid-seventeenth century, this principality was the domain of Prince Budjab (1577–1652). His descendants ruled the principality for fourteen generations until 1923. The map names sixty-nine cairns (*oboo*) that mark its border. It also shows the administrative center of the principality, and gives the height of the mountains. Courtesy of the National Central Archive of Mongolia.

authority over them, for they also swore their oath to Ögödei.[81]

SHM is clear on how this political authority works. The khan is "the *ejen* of the *ulus*"; what he does is "rule" (*medekü*) the *ulus*, and ruling makes one the *ejen*. In ruling, the khan issues commands/decrees (*jarliq*), gives laws (*jasaq*), and punishes those who break them.[82] The khan's command alone is a decree, and no one but the khan gives the laws. Likewise, if a law or decree is broken, it is the khan who ultimately decides who is to be put to death and who is to be punished: ultimately, ruling is a legitimate authority over life and liberty, and the control of life and liberty makes the ruler *ejen*/sovereign.[83] The khan also names (*nereyidükü*), appoints (*tüsikü*), and removes (*bawūlba*) all officers, beginning with commanders.[84]

Ruling is hierarchically nested, and commanders rule, too. In ruling, the khan "commands" (*noyalaqu/jasaqu*); likewise, commanders command their respective divisions.[85] There is also a division between "home" (*ger*) and "field" (*ke'er*): "Field matters should be ruled by those in the field, domestic matters should be ruled by those in the home."[86] For this reason, when Chinggis Khan dispatched Sübe'etei Ba'atur after Merkid in 1218, he gave Sübe'etei the authority to punish "command/law/decree breakers"; Sübe'etei was given the authority to put to death those who broke his commands.[87] This authority made Sübe'etei the commander, or ruler, in the field (*noyan*).

The khan owns the *ulus*, and the *ulus* is his patrimony. In this sense, the khan is not just the supreme ruler; he is also the supreme owner. Chinggis Khan apportions his subjects to his sons and brothers as their "share" (*qubi*) or "share property" (*emčü qubi*).[88] The khan also has the power to create new polities; Chinggis Khan creates *ulus*es for Jochi and Chaghadai, making them "princes who rule *ulus*."[89] In a word, the Chinggisid *ejen* is the ultimate sovereign, and *SHM* can be seen as spelling out the doctrine of khan's sovereignty.[90] If we adopt Weber's description of political authorities, the Chinggisid *ejen* is a patrimonial sovereign—one who owns the state as his personal patrimony—a ruler with "full proprietorship of the [state]," which Weber called "sultanism."[91] This is the *imperium* and *dominium* that appear under the heading of "Imperatoris Tartarorum seruile in omnes imperium" in a section titled "De potestate Imperatoris & Ducum eius," where John of Plano Carpini describes the "mirabile dominium" of the Mongol Great Khan.[92] And this is Jean Bodin's "absolute and perpetual power (*puissance/potestas*) to dispose of all possessions, persons, and the entire state at his pleasure, and then to leave it anyone he pleases, just as a proprietor can make a pure and simple gift of his goods for no other reason than his generosity."[93] "This power is absolute and sovereign," and it is the power of the "Great King of Tartary"—the epitome of sovereign power and the archetype upon which Bodin developed his concept of sovereignty.[94] In developing the concept of sovereignty, Bodin's aim was to legitimize the patrimonial authority of the French king. And it is the proprietary dimension of sovereignty that, during a subsequent era of colonialism, seems to have grounded the discourse of people-to-territory transformation.

This chapter has offered an alternative history of sovereignty as practiced by the Mongols. My aims have been twofold: to trace the convoluted history of the people-to-territory idea from the seventeenth to the twentieth century, and to argue that in fact there was no such transformation. As we have seen, the Great Divide is a figment of evolutionary materialist thinking, rooted in Westphalia but ultimately projected globally. Henry Maine's tribe sovereignty was a construct of colonial ideology; so was Lattimore's application of the same scheme to Mongolia. Yet in Inner Asia as elsewhere, the historical record shows that political authority has always had territorial expression and territorial limits. The actual transformations we find, on the steppe as elsewhere, cannot be captured by a simple binary. Instead of a sharp break between a "people-based" and a "territory-based" regime, we have seen more subtle shifts over time: from divine majesty to popular sovereignty; from traditional authority to legal authority; and from patrimonial-feudal government to the kind of republican federation embodied in the constitutions or *Rechtsstaat*'s rule of law. In a word, the Mongols did not need to wait for the Daiching to bring territoriality to them. On the contrary, the statecraft practices of the steppe—including cartography—coevolved with those of their coeval states across Eurasia.

FIGURE 3.1. *Map of the Traces of Yu* 禹跡圖, engraved on stone in 1136 in the State of Great Qi. The rubbing dates from circa 1800. 80 × 79 cm. Public domain. Available at the Harvard University Rubel Fine Arts Visual collection, http://id.lib.harvard.edu/alma/990100927990203941/catalog.

3

Peter K. Bol

Maps for Failed States

This is a story of two maps, created five hundred years apart. The first, from 1136, on a scale varying from 1:4.5 million to 3.5 million, is devoted largely to the territory under the civil administration at the time. The second, from 1640, on a scale of about 1:25,000, shows one of the eight townships of one of the about 1,300 county-level administrative units in one of 140 prefectures in one of 15 provinces. I read the depictions of territory on these two maps as articulating two quite different political aspirations. The first map reveals an aspiration to empire, but an empire whose geographical parameters have their origins in and are justified by antiquity. The second map populates the local landscape with villages named by lineages (families that maintain continuity over many generations) and aspires to an alternative to the state's centralized bureaucratic system. Calling these "maps for failed states" comes from recognizing that the aspirations of each of the two maps were not realized—for better or for worse, depending on one's point of view.

The *Traces of Yu* from 1136

The *Map of the Traces of Yu* 禹跡圖, engraved in 1136 on a square stele, replicates a map from 1100 done

in the city of Chang'an (modern Xian in Shaanxi) (figure 3.1). The geography of the map is earlier, as it reflects the administrative geography of the Song dynasty in 1087–89.[1] It locates mountains with a name, rivers with lines of varying width, and lakes with polygons. It locates prefectures by name but leaves out counties. Its spatial accuracy is superior to all other maps known from the time. Although the use of a grid was not new, this is the earliest extant gridmap. It has been held up as an example of something in China's past that anticipated modern methods, but for the map makers the point was the connection to antiquity.

Traces of Yu refers to King Yu, the third (after Yao and Shun) of the three legendary sage kings of antiquity who created a government that unified the realm. Yu, having been charged by Shun to deal with the great floods, traversed the land, channeling the rivers as need be. Out of these years of travel and water management came a report on the geography, products, taxes, and natures of the peoples of nine named regions. His report, known as the *Tribute of Yu* 禹貢, is included in the *Book of Documents* (one of the Confucian Classics). It is China's most ancient geographical text. Yu may have been legendary, but the text was not, although it was certainly from much later. It describes the diverse geography, tax tribute, and customs of nine regions in ways that are largely recognizable. The territory covered by the map goes beyond that described by the *Tribute of Yu* in the south but includes all territory of the Northern Song dynasty (960–1126). The map is not based directly on surveying; however, locations on the north–south axis suggest observation of the sun's angle from a substantial number of sites.[2]

Maps are propositions, but what are the propositions here?[3] The first is that the present is continuous with the beginning of civilization in antiquity as represented by the sagely King Yu. The cartouche on the map states that it includes ancient and modern names of rivers and mountains. It is not meant to be a map of all that Yu reported; rather, it roots the present in antiquity, at the beginning of (Confucian) civilization, by showing through its plotting and naming of rivers and mountains that the prefectural seats of the present exist in a physical landscape that is ancient. *Traces* was unique in using a map as its medium, not as its goal.

A contemporary geographical text connects the present to antiquity and history more effectively than the map. This is the *Extensive Record of the Realm*, a privately compiled historical gazetteer in two parts. The first part gives the slices in time, the upper administrative hierarchy at various points from antiquity to modern times (the *Tribute of Yu*, the Seven Warring States, Qin, Han, the Three Kingdoms, Jin, Tang, Tang military governors, and the Five Dynasties), but lists within each slice the Northern Song prefectures. The second part is a time series; it gives the Song administrative hierarchy, down to the county (*xian*) level, and notes the administrative history of each unit from its establishment in the past to the present. This was meant to be more than a reference work. The preface suggests that when readers see the spatial continuity between the present and antiquity, they will agree with the compiler's stated opposition to the court's efforts to expand Song territory beyond what history allows.[4]

Both works—one cartographic and the other a kind of database—construct for the reader a larger entity that is continuous over time, one by focusing on physical geography and the other on administrative geography. Similar efforts to link the present to the past or to antiquity were taking place in other spheres as well.[5] The map is of a piece with these efforts.

But there are two things the map does not do that other maps at the time were doing, which by their absence from the *Traces of Yu* might be taken positively as propositions. First, it shows a world without a "Great Wall," a fortification dividing (at least symbolically) the sedentary taxpayers on the south side from the pastoral tribes on the north. There were various extant walls on the northern frontier in Song, and they figured in conceptions of the Song state.[6] A commercial historical atlas from the period, the *Handy Maps through the Ages*, uses a template with a single wall running east to west, even quite anachronistically for its map of the *Tribute of Yu*.[7] Second, *Traces* does not recognize the existence of the states or tribes whose territory fell in part or whole within the map's grid. At the time of the map's original creation in 1080s, this meant Song itself (although prefectural seats are noted), the Tanguts' Xixia in the northwest, Khitan's Liao in the north, Dali in the southwest, and Dai Viet in the far south. It does not mark off the sixteen prefectures that had been part of the Tang dynasty (619–907) and that the Liao had occupied by 938, well before the Song founding in 960, although Song assumed they were its by right and had waged wars in vain to recover them. Song still thought to claim them, however, as the *Handy Maps* does on its map for 1085. Second, it does not show the existence of the various tribal peoples—foreign states, in fact—on the frontiers, a matter that the *Handy Guide* addresses with a separate map entitled *Map of the Civilized and the Tribes* 華夷圖. I will return to this later.

Traces thus includes territory beyond Song administrative control. It can be thought of as delineating what in Chinese was known as "all under heaven" (*tianxia* 天下). Song conceived of itself as the "central country" (*zhong guo* 中國) surrounded by tribal peoples (*yi di* 夷狄) but aspired to a world where all under heaven and the central country were one and the same. This did not reflect the facts on the ground; however, that may be irrelevant if we consider the map as aspiring to a world without borders and free of the hostilities necessarily brought into existence by foreign others.

To this point I have discussed the context in which *Traces* was originally drawn, though the 1136 maps did not at all suggest it was a product of the Song dynasty. Why, then, engrave the map in 1136? The answer is because it was a map meant for a new

state. The date I have translated as 1136 on the map's cartouche reads: "Engraved in stone in the fourth month of the seventh year of Fuchang." *Fuchang* is the name of the reign period of Liu Yu 劉豫 (1073–1143), emperor of the State of Great Qi 大齊國. A brief account of the geopolitical situation will lay the ground for my calling this a map for a failed state. The Song, in hope of taking the sixteen northern prefectures, had made an alliance with the Jurchens, in the northeast territory of Khitan's State of Great Liao 大遼國, who had rebelled against Liao and proclaimed a dynasty of their own, the State of Great Jin 大金國. The war was successful, but Song went back on its commitments and the Jurchens marched on the Song capital of Kaifeng, captured the emperor and his father the former emperor, and went back north. However, they were not prepared to take control over north China and sought a proxy instead. First, they agreed with the former Song chief councilor, Zhang Bangchang 張邦昌 (1081–1127), that he would create the State of Great Chu 大楚國 with a capital at Nanjing on the Yangzi River and be its emperor. The goal apparently was to set up a state in the North China Plain that would be capable of paying the subsidies to Jin that Song, now in the south, had once paid to Liao. However, within a year Zhang gave up his claim and "returned" to Song, where he was promptly executed as a traitor. This left Jin looking for another ally, one capable of bringing order to the northern plain. This was Liu Yu.

Liu's biography in the official history of Song begins from the assumption that Liu was illegitimate, a "traitor." But the biography is quite detailed about events in Great Qi, for it also serves as a history of that short-lived state as well as being a biography of Liu himself.[8] Liu, the first in his family to pass the civil service examination, had a very respectable career, serving as vice director at the national university in 1124 and then as judicial intendent of Hebei when he was told to take appointment as the prefect of Jinan in Shandong Province in 1127. Seeing risings in the area, Liu asked the Song court (such as it was) to appoint him to a southern post. The court refused, and he was left to defend Jinan. The Jin made him a better offer in 1128/3: he could take control of everything south of the old course of the Yellow River. Two years later, in 1130/7, they recognized him as an emperor of his own State of Great Qi, to occupy the land between Jin in the north and Song in the south.

The Song history gives a litany of accounts of prefectural seats won and lost in the war, but along the way it notes events that lead inescapably to the conclusion that at the time, Liu was successfully establishing himself as a successor to the Song dynasty, which had, after all, lost its emperors and its capital, and which had not yet solidified its position in the south. In short order, Liu put together a court in the Song style: he appointed high officials, settled on a capital, called on his officials to submit memorials of frank advice and criticism, made his mother the empress dowager, appointed an empress, and announced that the next year (1131) would be the first year of the Fuchang ("Great Prosperity") reign period. He was acting like a "son of heaven"; he aimed to be an emperor, not a puppet. By the end of 1131, his aspirations were clear: he began attacking Song in the south. In 1132, he moved his capital in Kaifeng (the former Song capital), honored his ancestors in the Grand Temple there, offered the seasonal sacrifices, promised government reforms (no more eunuchs; no selling of ordination certificates; promotion by merit, not seniority), and managed over 100,000 troops in thirteen armies. In 1133, he acquired a navy of sixty warships. In 1134, he restored the civil service examinations, issued an imperial edict asserting his intention to "unify heaven and earth and all those in the four directions" 混一六合, and campaigned against Song. In 1135, he sent his Jin allies a map of sea routes and a model for an oceangoing ship, urging them to join his campaign by sea. On schedule, the civil service examination was held again in 1137. But the Jin came to see him as a danger and turned against him; he could not deal with enemies on both fronts.

The engraving in stone of the *Traces of Yu* in 1136 made possible the distribution of the map to the schools training examination candidates through reproduction by rubbings. It was of a piece with Liu's very traditional approach to imperial state building. He was putting his reign title on an existing map— the best national map anyone had ever seen—and using it for his own purposes. It depicted a vision of the empire his regime wished to rule.

But my reason for calling this a map of a failed state is not only because Liu Yu failed in his ambition. The idea that the empire should be based on territory continuous with its ancient history and that this constituted the traditional area of the "central country" survived; but the idea that this area should limit the territorial reach of the state was rejected from then until the present for all but 276 years. And this had everything to do with what the map did not show: the surrounding foreign peoples and polities. Liu Yu's government was not unaware of this context, for six months after engraving the *Traces*, another was engraved on the obverse side of the stele, titled *Map of the Civilized and the Tribes* 華夷圖 (figure 3.2), another map originally drawn in the Song period.[9] Here the proposition was clear: we represent the civilized center surrounded by foreign others, as described in texts surrounding the map proper.

But in fact, those foreign others gave us the map of China today. The Jurchens' Great Jin dynasty was one of those others; after it had deposed Great Qi, it prepared to invade the Song in the south (it would be defeated in the great naval battle at Caishi in 1161). Jin territory extended far into the north. Then the Mongols conquered first Jin and then Song, including them in a vast empire, one that stretched into the Christian and the Islamic lands in the west. The Mongol rulers of that empire claimed they had what they had by God's gift to them; they accepted no constraints from antiquity and history. The Ming dynasty did drive out the Mongols in the fourteenth century and largely accepted the borders as justified by antiquity, but it became obsessed with creating a real continuous great wall to divide their civilized realm from the tribal others, only to be conquered by the Manchus' Qing dynasty, which like the Mongols incorporated Ming territory into a much larger territory. Jurchens, Mongols, and Manchus did not accept that antiquity and history could dictate the limits of their empire. With the exception of outer Mongo-

lia, now the Mongolian People's Republic, most of the Manchus' empire was taken over by the People's Republic of China.

The Map of Chongde Township from 1640

The map of Chongde township (figure 3.3) is one of eight township maps in the *Gazetteer of Yiwu County* from 1640.[10]

Township maps are not found in the contemporary gazetteers for the other seven counties in Jinhua prefecture, although much of the information on the Chongde map appears in lists elsewhere in this gazetteer and similar information is found in the gazetteers of other counties. Yiwu in 1640 alone combined different kinds of maps. The compilers explain the symbolization (obvious enough that it does not warrant discussion here)[11] and why a gazetteer should begin with maps. Maps, they tell readers, were valued in both antiquity and later imperial history. The administrative geographies of the Tang, Song, and their own Ming dynasty included maps. But more important, maps and local information should be valued because "all-under-heaven is an accumulation from the individual county 天下者，一縣之積也." They also state that these township maps are particularly detailed and "greatly add what earlier gazetteers did not have 大增前志之所未有."[12] The Yiwu county gazetteer only began to include maps in 1353, but no edition before 1640 is extant.[13]

Most elements on the map are discussed or referenced in the gazetteer text. A few examples will suffice. The most famous of all the tombs was that

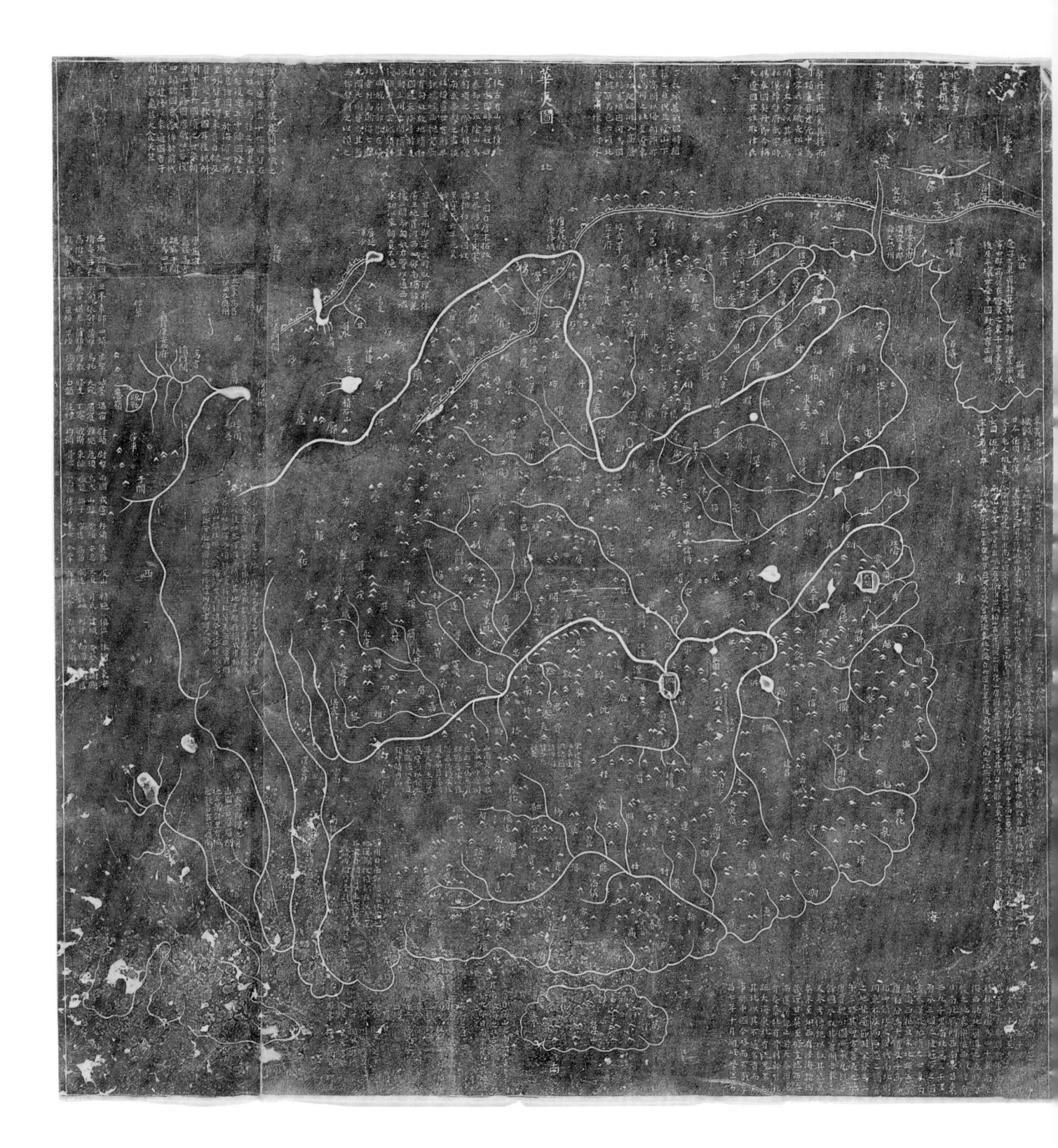

華夷圖

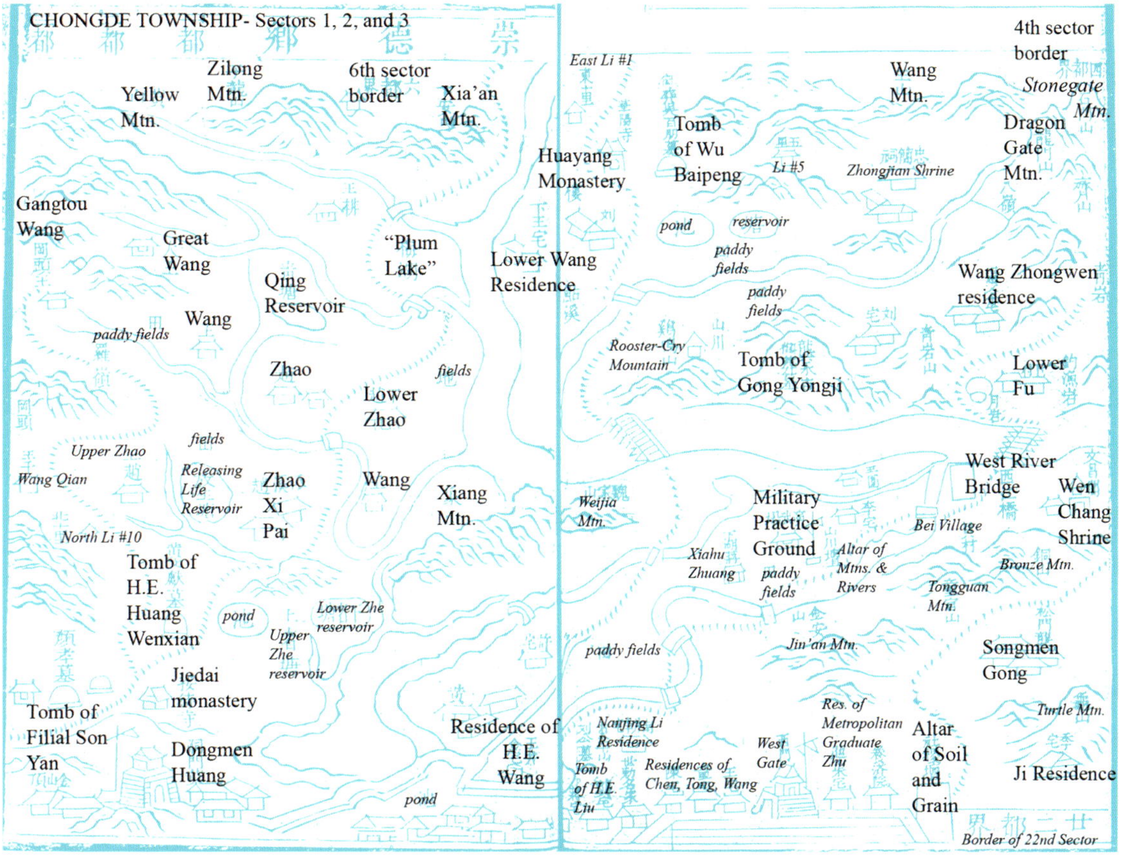

The map labels (left to right, top to bottom) read:

CHONGDE TOWNSHIP- Sectors 1, 2, and 3

4th sector border

East Li #1

Wang Mtn.

Stonegate Mtn.

Zilong Mtn.

6th sector border

Xia'an Mtn.

Yellow Mtn.

Tomb of Wu Baipeng

Dragon Gate Mtn.

Huayang Monastery

Zhongtian Shrine

Li #5

Gangtou Wang

Great Wang

"Plum Lake"

Lower Wang Residence

pond   *reservoir*

paddy fields

Wang Zhongwen residence

Qing Reservoir

paddy fields

Wang

paddy fields

Rooster-Cry Mountain

Lower Fu

Zhao

fields

Tomb of Gong Yongji

Lower Zhao

fields

West River Bridge

Wen Chang Shrine

Upper Zhao

Releasing Life Reservoir

Zhao Xi Pai

Wang

Xiang Mtn.

Weijia Mtn.

Military Practice Ground

Bei Village

Bronze Mtn.

Wang Qian

Xiahu Zhuang

Altar of Mtns. & Rivers

paddy fields

Tongguan Mtn.

North Li #10

Tomb of H.E. Huang Wenxian

pond

Lower Zhe reservoir

Upper Zhe reservoir

paddy fields

Jin'an Mtn.

Songmen Gong

Turtle Mtn.

Jiedai monastery

Res. of Metropolitan Graduate Zhu

Tomb of Filial Son Yan

Dongmen Huang

Residence of H.E. Wang

Nanjing Li Residence

West Gate

Altar of Soil and Grain

Ji Residence

Tomb of H.E. Liu

Residences of Chen, Tong, Wang

pond

Border of 22nd Sector

FIGURE 3.3. *Map of Chongde Township, Yiwu County, Jinhua Prefecture, Zhejiang* 金華府義烏縣崇德鄉, from the *Yiwu County Gazetteer* 義烏縣志 of 1640. The map shows various landscape features and villages. Here villages are identified by family surnames such as Zhao, Wang, and Fu. Public domain.

FIGURE 3.2. *Map of the Civilized and the Tribes* 華夷圖, engraved on stone in 1136 in the State of Great Qi on the obverse of the *Traces of Yu* 禹跡圖. Around the map of the Chinese area are descriptions of the surrounding foreign peoples. In contrast to the *Traces of Yu*, it depicts a Great Wall, although it is largely imaginary. Public domain.

of the filial son Yan, whose filial piety two millennia previous had inspired the name of the county. The gazetteer tells of the shrine that was built next to the grave; the officials who, beginning in the thirteenth century, honored him; the creation of a land endowment in 1441 to pay for the ceremonies and sacrifices; the charge to the monks of the nearby Jiedai Monastery to carry out the ceremonies; the texts of the commemorative inscriptions; and so on. There are biographies of the other figures whose tombs are on the map. A list of religious sites includes a history of the Huayang Monastery. The section listing the mountains in Yiwu states that Dragon Gate Mountain is the tallest in the county. The bridges sections notes when West River Bridge was built and how it was paid for. We can find information about the current and previous location of the military practice ground, and so on.

That the gazetteer supplies such details is not surprising. In 1418, the court announced rules for the contents of gazetteers. These included administrative units, walls and moats, mountains and rivers, local products, military sites, government buildings, monasteries and temples, bridges, and historical sites (e.g., tombs). The rules asked for more than this map shows: population numbers, taxes, schools, lists of local officials, biographies of local personages, and literary works pertaining to the place.[14] However, in Yiwu and elsewhere, local compilers added and elaborated. The elaborations were of a piece, greatly increasing the information on local literati, providing lists of those involved in the examination system, and writing biographies of local figures organized by category (noteworthy official, writer, Confucian scholar, etc.). A concern with remembering locals of renown is evident from the tombs that appear on the map.

But this map, like all the other township maps, includes another feature that is noted but not discussed in the gazetteer. These are the villages, given on the map by house icons. Villages in China have names, typically of landscape features—for example, a nineteenth gazetteer map from the neighboring county of Yongkang reflects the actual territorial dimensions of the county and names all the villages along routes radiating out from the county seat, and almost all the names are landscape features.[15] The same is true of the 1894 map of Yiwu county, done with modern mapmaking techniques.[16]

But the Chongde township map does not name the villages it includes according to their landscape-style names. Instead, it gives a surname (places called "residences" are also villages). The 1640 township maps which identify villages by family name are replicated exactly, using the same woodblocks, in the editions of 1692 and 1727. And then, in 1799, all the township maps disappear. The compilers explain that the situation in the townships has changed so much that they have replaced them with drawings of scenic sites. For villages, they refer readers to the section on administrative subdivisions. There we do find under Chongde township a list of villages for each of the three sectors. The village names are the landscape names we find in the 1894 map, but under each is a family name, letting us know that Willow Village is of the Gong family. The fact of villages being connected to families has not ceased, just the mapping of them. So what?

Anthropologists and social historians have studied lineages extensively because in southern China these were and still are ubiquitous, constituting a form of "bottom-up" social organization that existed independently of the government's administrative system. Lineage organization was where the state and the rural population met; it was their common language.[17] Lineage villages are known colloquially as "single-surname villages" because practically all the male inhabitants are descended from a common ancestor, with daughters marrying out and wives marrying in due to the taboo on same-surname marriage. The village is centered on an "ancestor hall" for the worship of the ancestors (and sometimes halls for the several branches of the lineage). In Yiwu and the surrounding area, lineage villages appear to have solidified around the fourteenth century.[18] Above all, they relied on written genealogies to keep track of membership, a necessity since over tens of generations there might be more than two thousand people in a single generation spread over several villages and the birth years within the same generation could differ by two hundred years.[19] According to their genealogies, the formation of these lineages began with literati families. Leading scholars at the time formulated an ideology of the lineage as the foundation for individual and social morality, as a source of shared historical consciousness, and as the way that literati could bring order to society from the bottom up without holding government office.[20]

This last point is important because it puts the lineage in competition with government. Here I quote Fang Xiaoru 方孝孺 (1357–1402), the favored student of Song Lian 宋濂 (1310–1381), the most influential local scholar from the fourteenth century:

Literati can transform the world without office by harmonizing their lineages. How is it that we can transform such a large world by harmonizing our lineages?

People all want to harmonize their lineages; the problem is they do not know how to do it correctly. If we go first, then who will not follow us, and, when all lineages are harmonized, then who in the world will join in evil? If the evil cannot take license, then perfect order is but a step away.

There are three ways to harmonize the lineage: do a genealogy to connect the lineage, visit the grave of the first resident ancestor to bind the hearts [of members], and strengthen the rites of kinship to nurture kindness to others. . . .

When all three are practiced, then even a [mere] literatus can transform others; how much more so one with office. Changing all society's customs is not difficult; how much less so in the case of something as near at hand as the community and village. For those who are near[,] it ought to be easy to act; for those with office, it ought to be easy to transform them. That none act and that none transform is because the Way is known to only a few. If those who know the Way are also those with office, how could people not look up to them?

The Jinhua Songs are the lineage of the historian Song [Lian], who in morality and literary learning was teacher to the age and who in the practice of the Way was ahead of his lineage. All that could be done to harmonize the lineage he did. This is his genealogy. It was not just of his one generation; it will be kept by later generations as well. Nor was it a matter of his immediate family; it was adopted by the whole lineage. If everyone of the lineage in the future could have values like his, then they could make do without [a genealogy], but, if they do not, then even with a model through which [his values] would be transmitted, one would still be afraid [that his values] would be lost; how much more so a mere genealogy. On this account I have taken it on myself to append his views to the end so that people in the future might draw on them. Going from one lineage to extend to the whole world must begin right here. That was Song Lian's point.[21]

The line from the fourteenth century to the present is direct. Although lineages became segmented and their relative wealth and power shifted over time, the structure has remained. The map simply recognized the reality on the ground and drew attention to it.

From the map the ideological context is not apparent, and from the gazetteer there is no explanation of how these single-surname villages came to be. We do know, however, that the policies of the early Ming dynasty (1368–1644) sought to restore social stability after decades of rebellion and civil war at the end of the Mongols' Yuan dynasty. The Ming founder envisioned villages as self-sustaining moral and economic entities and granted to their leaders considerable au-

thority outside the bureaucratic hierarchy, although at this point, as single-surname villages were just emerging, this fact apparently did not enter into the government's calculus. The founding emperor was suspicious of his officials and autocratic in relation to his bureaucracy, but not in his social policy. He purged and executed many officials, including Song Lian, who had been serving as a historian and intellectual advisor. For the sins of his descendants, Song Lian was exiled to the west and either committed suicide or was killed en route. Fang Xiaoru, his student, remained in good standing—the hope of the literati, it is said—and became a senior advisor to the next emperor. His writings make clear that he envisioned a world where power devolved to localities and local literati lineages. His career was cut short in 1403 by the Yongle emperor's usurpation, which Fang refused to support. He was executed, along with his kin to the ninth degree and friends and allies. The new regime kept the local institutions but moved to reassert central authority at the expense of local autonomy. But by the late 1500s, the Ming system was in disarray: an influx of silver from Japan and the Americas was heating up the commercial economy of the southeast, regional disparities were becoming extreme, inequality was increasing, social unrest was spreading, and the rural institutions of the early Ming were no longer effective.

Fang Xiaoru was posthumously rehabilitated in 1584, descendants of Fang and other condemned men were released from control, and Fang's writings were republished.[22] His recognition fit with increasing local activism and calls for greater local autonomy as a way of coping with the failures of central authority. An old debate in political thought was once again very much on the minds of statecraft thinkers of the seventeenth century, particularly in the southeast.

This debate was framed as a choice between two models of government: (1) the centralized bureaucratic system, known as the "commandery and county" 郡縣 system that had been instituted when Qin brought all the independent states under its control in 221 BCE, and (2) the "enfeoffment" 封建 (sometimes translated as "feudal") system of antiquity. In the centralized bureaucratic system, the court sent out officials to administer the prefectures and counties, rotating them every two to three years to ensure that they would be more responsive to the court than local interests. Confucians faulted the short-lived Qin unification for its burying scholars, burning books, and abandoning the ideals of antiquity, but the centralized bureaucratic system has remained the fundamental political structure until today. In the seventeenth century, some Confucian idealists argued for a return to enfeoffment on the grounds that decentralization and the devolution of power to localities would ensure greater effort and freedom of initiative to alleviate the social problems of the day. Local officials would not be bureaucrats anxious for promotion to office at court, but families who served as local officials over many generations and were devoted to the well-being of the locale.[23]

However, there were different ways of thinking about what the devolution of power to the locality would mean. One called for radical intervention in local society. Its proponents insisted that enfeoff-ment required land redistribution to ensure that all households had the land necessary to support themselves and pay their taxes. This, they said, was the ideal model bequeathed by antiquity. Decentralization meant that the state would have far greater power over daily life than had been the case. The idea of land redistribution, which involved confiscating the land of landowners and, in effect, making farmers the tenants of the state, would not be realized until the 1950s.

The alternative was to recognize that the lineage village was the dominant social formation in the countryside. This offered a different way of thinking about enfeoffment, one that Fang Xiaoru had articulated. In this view, local society would be dominated by self-supervising lineages that would see to needs of their members, keep the peace, and cooperate with the state. By the 1600s, lineages in places that had not been destroyed by natural disasters or violence had seen seven or eight generations of growth. When the Chongde township map was made, in 1640, the Ming dynasty was facing economic and military crises; it was exactly at such a moment when the central government, riven by factionalism, was incapable of dealing with domestic rebellion and foreign threat, that Fang's view of the role of lineages promised a possible way of enlisting the population and saving the state. It is perhaps an indication of a growing acceptance of Fang's views and a disavowal of the centralizing policies of the ruler who executed him that when the Ming court fled south after the Manchus conquered Beijing in 1644, it honored Fang and recognized him as a model by granting the posthumous title of Wenzheng 文正, "Cultured and Correct."

Nevertheless, even if the state and the lineage generally found it necessary to accommodate each other, the relationship was not always harmonious in theory or in practice. Local officials would have preferred not to accommodate the power of lineages. The 1640 map recognized for all to see the actual situation on the ground, which remained the case even when this graphic recognition was replaced by bland landscape pictures. A country whose landscape was dominated by lineage villages was not the way defenders of the centralized bureaucratic system wanted to conceive of the realm; lineage power was suffered by the government, not treated as an opportunity to rethink the relationship between society and state.

One of the conundrums of later imperial China in the south is why the imperial state survived in its centralized form, given its small size and the power of lineages. Another map from 1640 offers an explanation. This is a map of the entire county (figure 3.4), at the center of which are two institutions that existed in every county: the county administration and the Confucian school.

The school system and the civil service examination system it fed provided a merit-based venue that crossed lineage boundaries, brought literati into prefecture-wide social networks, gave them as a group influence over local government, and guaranteed that some of them would enter government service. The governmental institutional system could stay small as population doubled and quadrupled be-

cause much power had devolved in fact to lineages. But lineages were connected to the world of literati and government through their ancestors, and genealogies contained biographies and patents of office that drew attention to the kin, however rare, who had passed the examinations or held office.

Yet the *Map of Chongde Township* deserves to be called a failed map because the government was not, and is not, willing to rethink the relationship between state and family. During the twentieth century, both Nationalist and Communist governments were determined to increase state power in pursuit of modernization. They saw localism and lineages as obstacles and set out to diminish and destroy them. But the story is not over. After 1949, the compilation of local gazetteers and genealogies was forbidden. And yet, in the 1980s, both began to reappear—gazetteers by government mandate and genealogies in their wake by private initiative, although local officials were often reluctant to allow their printing. As the author of a preface to one of these resurrected genealogies has written, now that the Cultural Revolution with its attack on tradition and family is over, prosperity is returning; so we turn again to writing local gazetteers and genealogies.[24]

Guojia, "state" or "nation," combines two terms, *guo* 國, "country" or "state," and *jia* 家, "family."[25]

Whether or not the reader agrees that 1136 and 1540 are maps of failures, it is certain that the first map is about state building and the second one is about family. They each raise a fundamental question: What is the legitimate extent of the state and

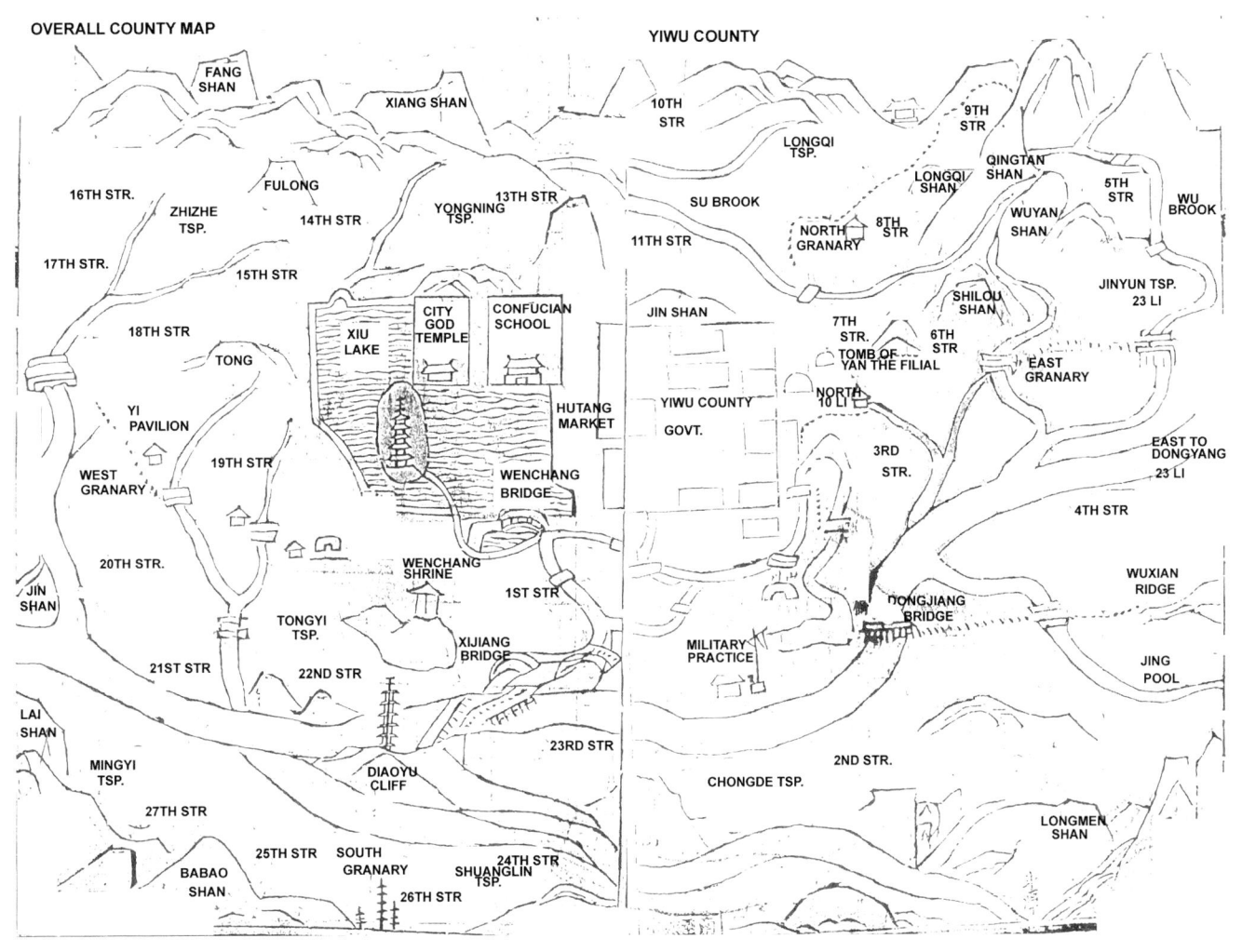

FIGURE 3.4. *Map of Yiwu County, Jinhua Prefecture, Zhejiang* 金華府義烏縣, from the *Yiwu County Gazetteer* 義烏縣志 of 1640. The map shows the entire county distorted into a square, with the county offices and Confucian school at the center. Public domain.

of the family? Both maps are representations of answers to those questions. In 1136, the territory of the state that Liu Yu's Great Qi aspired to went somewhat beyond the borders of Song, but it was justified by a claim to a history going back to the very beginning of unified empire in antiquity. In 1640, the temporal extent of the family was open; the lineage had a starting point in the "first migrant ancestor," but no end point. However, the map is spatially constrained by the existence of other lineages, not unlike the second

map on the 1636 stele which acknowledged that the territory Great Qi assumed to be historically theirs was surrounded by alien peoples. I think it is fair to say that both propositions have been contested if not rejected. The visions of empire of the Jurchens' Jin, the Mongols' Yuan, the Manchu's Qing, and the People's Republic of China do not accept that the *Traces of Yu* suggests limits on the extent of the state. Since the twentieth century, the centralizing and modernizing Chinese state has denied the autonomy of lineages, and in fact has at times sought to destroy them. And yet both situations persist today. The peoples on the northern, western, and southwestern reaches have different languages, writing systems, and political histories; they have not willingly accepted forced assimilation. And in the southern countryside, lineage villages have not disappeared, but have been updating their genealogies and reasserting their economic and social interests.

Ali Yaycıoğlu

On the Ottoman Arguments during the Congress of Karlowitz (1699)

An anonymous oil painting housed in Museo di Palazzo Poggi (figure 4.1) depicts a scene from the Karlowitz Congress (1698/99). This diplomatic summit ended a fifteen-year war between the Ottoman Empire and the Sacra Lega, a Christian alliance between the Habsburg Empire, the Republic of Venice, the Polish–Lithuanian Commonwealth, and the Tsardom of Russia. The congress took place in Sremski Karlovci (in today's Serbia) from October 1698 to January 1699. Like the famous summit of Westphalia fifty years earlier, Karlowitz featured multilateral diplomacy.[1] Yet this was the first time such a diplomatic assembly included the Ottoman Empire.[2] The painting dramatizes this cultural encounter.

At the center of the painting, conspicuous in white turban and red kaftan, is Mehmed Rami Efendi (1655–1708), the plenipotentiary of the Ottoman state. On his right is Alexander Mavrocordato (1641–1709), the Ottoman Greek dragoman (interpreter). Framing the two Ottoman representatives are their counterparts from the Habsburg Empire: at left, the cartographer Luigi Ferdinando Marsili (1658–1730), and at right, a figure who probably represents Count Leopold Schlick (1663–1723). A man with a large wig in the background is likely Carlo Ruzzini (1653–1735), the Venetian plenipotentiary. Three scribes—seated at the left and right edges of the painting and behind Mehmed Rami Efendi—record the proceedings. The shadowy figure watching from far right may be William Paget (1637–1713), the British ambassador, who mediated throughout the congress.

In the foreground, the artist has painted a map. Rami is pointing toward the prominently featured river, while Marsili points in a cardinal direction from the gathering place, presumably toward the spot Rami is indicating on the map. Mavrocordato and Schlick are in dialogue while the scribes are writing down the minutes.

Did the Ottomans negotiate new borders with the Austrians using a map of this kind? Marsili had prepared several maps for the Austrian and Venetian

For Dariusz Kołodziejczyk. I would like to express my gratitude to Serhat Aslaner, Abdurrahaman Atçıl, Ercan Akyol, Lane Baker, Rowan Dorin, Cristina Carile, Özer Ergenç, Ali Aydın Karamustafa, and Murat Somer for their comments and help throughout the journey of this article. Special thanks to Kären Wigen and Nora Barakat for close reading of the chapter, making superb comments, and helping me to shorten it.

mission in advance of the negotiations. In fact, the Karlowitz Congress was one of the earliest peace conferences in Europe in which cartographic tools were used. The Austrian and Venetian missions planned to start negotiations on the new border lines with documents based on Marsili's maps, proposing that a team of Ottoman and European surveyors be sent to the region to fix the borders on the ground in accordance with where the negotiators marked them on the map. But the Ottoman mission disagreed. Arguing that maps like Marsili's did not reflect local conditions, they insisted that borders could not be decided from a distance. Instead, they proposed that a multilateral team visit each border zone without a predetermined chart or map. Details would be de-

termined on the ground, in consultation with local communities.

These competing stances reflect two different ways of envisioning the proper use of cartography in delimiting sovereign space. The Austrian–Venetian mission assumed that maps agreed on at a diplomatic table constitute authoritative representations of space. The Ottoman mission, on the other hand, countered that the geographic information crucial to boundary making could not be ascertained in this way. While the congress could agree on principles, borders should be delineated on the ground, with input from local notables. The particular prevailed over the general; local knowledge could not be dismissed.

Context

The Karlowitz Congress marked a turning point in Ottoman history. After their armies suffered a decisive defeat at Vienna in 1683,[3] the Ottoman war with the Habsburgs had expanded on multiple fronts, from the Ukrainian steppes to Hungary, from Bosnia to the Peloponnese. In response, the great powers of central Europe in 1684 banded together to form the Sacra Lega: a Christian alliance whose members included Austria, Venice, Poland–Lithuania, and Russia, and whose goal was to end Ottoman expansion for good. Three years later, the Sacra Lega defeated the Ottoman forces in Mohács, ending 150 years of Ottoman rule in the Hungarian plains. This defeat triggered massive unrest in the Ottoman army and turmoil in Istanbul, putting an end to Mehmed IV's thirty-nine-year reign.[4]

In 1697, in the Battle of Zenta, the forces of the Sacra Lega defeated the Ottoman army again.[5] A considerable portion of the Ottoman officers were killed. The treasury to provision the janissaries was captured, as was the imperial seal. Ottoman Serbia and the Banat valley ended up in the hands of the Habsburg Empire.[6]

These wars are seen as a crossroads for the Ottoman Empire and the beginning of the Ottoman withdrawal from Europe and the Ottoman "decline."[7] Hungary, long a stronghold in Central Europe, was lost to the Habsburgs. On the Mediterranean front, the Republic of Venice avenged its own loss of Crete twenty years earlier by establishing control over the southern Adriatic Sea and the Peloponnese. The empire's military inferiority vis-à-vis the Habsburgs and its allies had to be recognized.[8]

By 1697, both the Ottoman Empire and the Habsburgs were ready to end the fifteen-year war. In both capitals, "peace parties" were consolidated among the ruling elites.[9] For reasons of their own, England and the Netherlands took the initiative to broker the peace.[10]

After several meetings with the British and Dutch arbitrators, the Ottoman Empire and the members of the Sacra Lega agreed on Karlowitz (Sremski Karlovci) in Serbia for the location of the peace talks. It was in this context that the Congress of Karlowitz convened in October 1698. The Ottoman delegation met with representatives of the Habsburg Empire, the Polish–Lithuanian Commonwealth, the Republic of Venice, and the Tsardom of Russia a total of thirty-six times. After tough negotiations, separate treaties were concluded with all parties.[11]

Uti Possidetis and Cartography

In a seminal work, Rifa'at Abou-El-Haj has declared the Karlowitz Congress a turning point for Ottoman boundary conception. With the resulting treaties, the empire's notion of fluid frontiers gave way to delimited and fixed borders. Karlowitz represented the formal closure of the Ottoman European frontier, which had been a zone of constant conquests and continuous expansion.[12] Gábor Ágoston argues that "the importance of Karlowitz lay not in the delimitation of the border but in the fact that the Ottomans accepted, albeit reluctantly, a European conception of sovereignty and territorial integrity of their neighbors."[13] There is truth in these views. As I suggest elsewhere, the Ottoman defeat in the long wars triggered a substantial, albeit gradual, change in Ottoman political thinking, turning an empire of conquest into an empire of defense, and in some ways an empire of peace. It also laid the foundations for a period of reform on various fronts throughout the eighteenth century.[14] But the border-setting process deserves a closer look.

Crucial here was a new international legal principle: *uti possidetis, ita possidetis* (as you possess [now], thus may you possess [going forward]).[15] This principle stipulates that each party should keep whatever territory it possessed at the conclusion of hostilities. In effect, the status quo should be legalized; all parties should withdraw claims over any lands controlled by their enemies at the conclusion of the conflict.[16] Originally a Roman property law concept, *uti possidetis* was refashioned as an international legal norm during these decades.[17]

Although such a principle was not totally new for the Ottomans, in most cases they were reluctant to withdraw their transborder claims.[18] During the preceding centuries of expansion, peace treaties had typically been seen as provisional. The empire had continued to support irregular warriors in the frontier to raid the border territories even after peace settlements. That the Karlowitz Treaty accentuated exclusive territorial integrity was an important step toward creating a more lasting border between the Ottoman Empire and the European states. The question was how to draw the lines.

By 1698, the application of *uti possidetis* had fostered a new use of cartographic tools. During the negotiations for the Peace of Westphalia in 1648, the Peace of the Pyrenees in 1659, and the treaties of Utrecht in 1713, maps were used as "supporting evidence for claims and as a means of agreeing to a particular division of authority."[19] In each case, border delimitation became a matter of verifying which locality was under the control of which power when the guns stopped firing. *De facto* control of territory, rather than the needs of people living in the disputed regions, became the central issue.[20] Drawing lines on maps in turn fostered abstract understandings of sovereign territories as homogeneous spaces without overlapping jurisdictions. These abstract spaces lent themselves to manipulation at a distance, even if the implementation of new divisions on the ground would not always be realistic.

Not infrequently, this process created havoc in borderland communities. Negotiators' maps might flagrantly disregard the economic and ecological circumstances of local life; a proposed boundary could cut off a village from its water sources or its agricultural hinterland. Other times, the mapmakers even intentionally targeted borderland groups—notably, pastoralists who routinely moved between states. At the end of the day, maps were technologies of power designed to serve states and their rulers, not local communities;[21] and multilateral congresses became platforms where claims expressed in cartographic form could be passed from one side to another at the negotiating table.

Marsili's Map Program

The Habsburgs came to Karlowitz with this mindset. Before the congress convened, several maps were prepared for the Habsburg and the Venetian delegations. According to John Stoye, "The Venetian ambassador Ruzzini describe[d] a conversation with Starhemberg early in August, when Starhemberg [a Habsburg General] referred to a collection of 'topographic graphical maps' (Ruzzini's own phrase) which he had recently commissioned from a number of engineers; these were to show the existing limits of Habsburg authority along the rivers Una, Sava, Danube, Tisza, and Maros. He hoped that the maps would help to clarify uncertainties and bring the two powers closer to agreement."[22]

The most detailed of the available maps were those prepared by Luigi Ferdinando Marsili, a Bolognese aristocrat.[23] After studying botany in Bologna, Marsili had developed an interest in geography and cartography. In 1679, he accompanied the Venetian officer Pietro Civran on a trip to Istanbul, during which Marsili met several Ottoman intellectuals and statesmen. In 1682, while serving in the Habsburg army as an officer during the Vienna campaign, he was captured by the Ottomans.[24] Following his release two years later, he returned to the front of the ongoing wars against the Ottomans in Buda and Zante.

In 1689 and 1690, Marsili prepared a set of maps illustrating the Ottoman–Habsburg borderlands. With the help of other cartographers, he later updated these maps, adding proposals concerning possible boundaries from the Adriatic coasts all the way to Transylvania. Marsili's maps would play a major role in the Karlowitz negotiations. Thanks to his knowledge of the Ottoman–Habsburg and Ottoman–Venetian frontiers, he was appointed as *consigliere* in the Habsburg mission.[25]

Marsili's cartographic projects and the reports he prepared before the Congress of Karlowitz provided the Austrian and Venetian mission with detailed cartographic information about the borderlands.[26] His cartographic program included large-scale maps, topographical illustrations, city views, charts, and plans.[27] While some of these illustrations were prepared with new cartographic techniques, others were more like panoramic drawings, reflecting the transitional nature of mapmaking in this period. One document in the Marsili archive (no. 51), dated between April and August 1698, lists a series of problematic borderlands, using maps to explain them.[28] The first three sections discuss Wallachia, Transylvania, and Belgrade in turn. A fourth section returns to Transylvania to consider the use of mountains and rivers to demarcate boundaries there, since artificial boundaries in the region would be difficult.[29] Another document (no. 58), dated April 1698, discusses the Una and Sava Rivers as natural boundaries between the Ottoman and Habsburg Empires in Croatia and Bosnia, itemizing fortresses under Ottoman and Austrian control.[30] In yet another report, Marsili prepared tens of maps and plans of fortresses and their topography on the Ottoman–Habsburg borders (figures 4.2a, 4.2b).[31] In accordance with the principle of

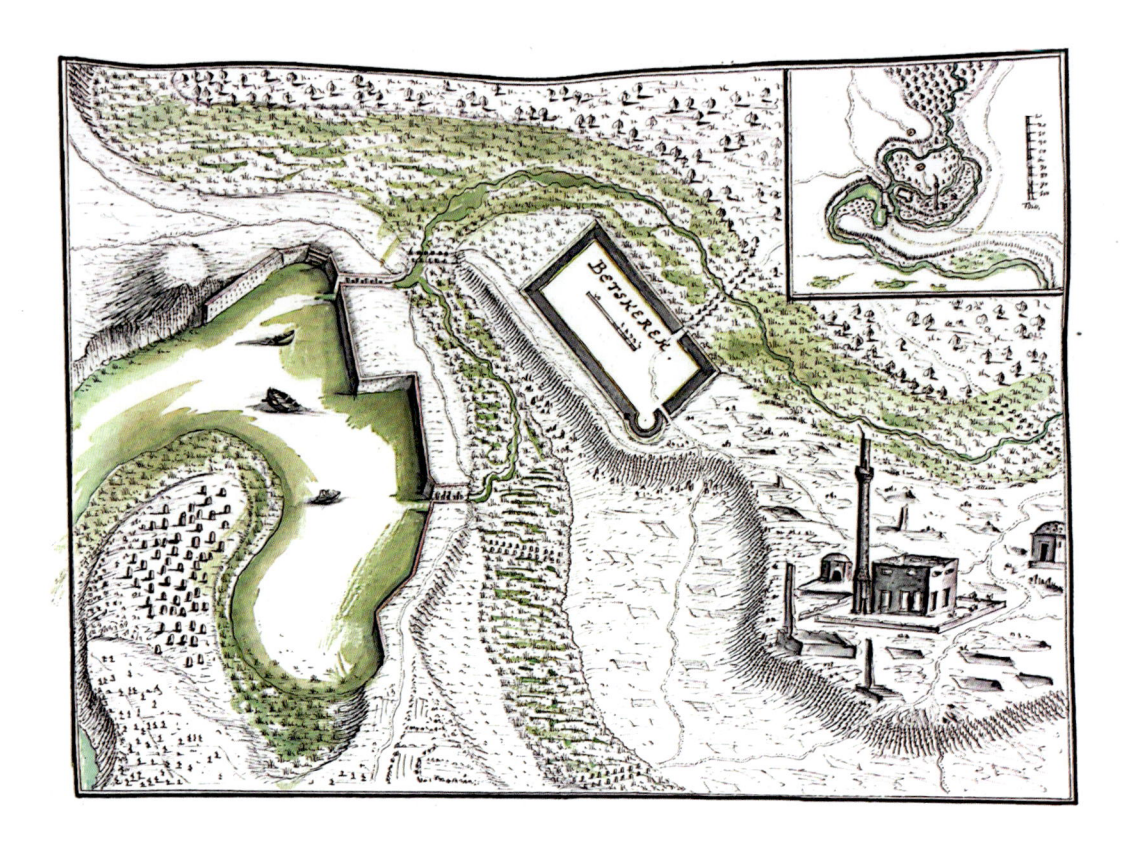

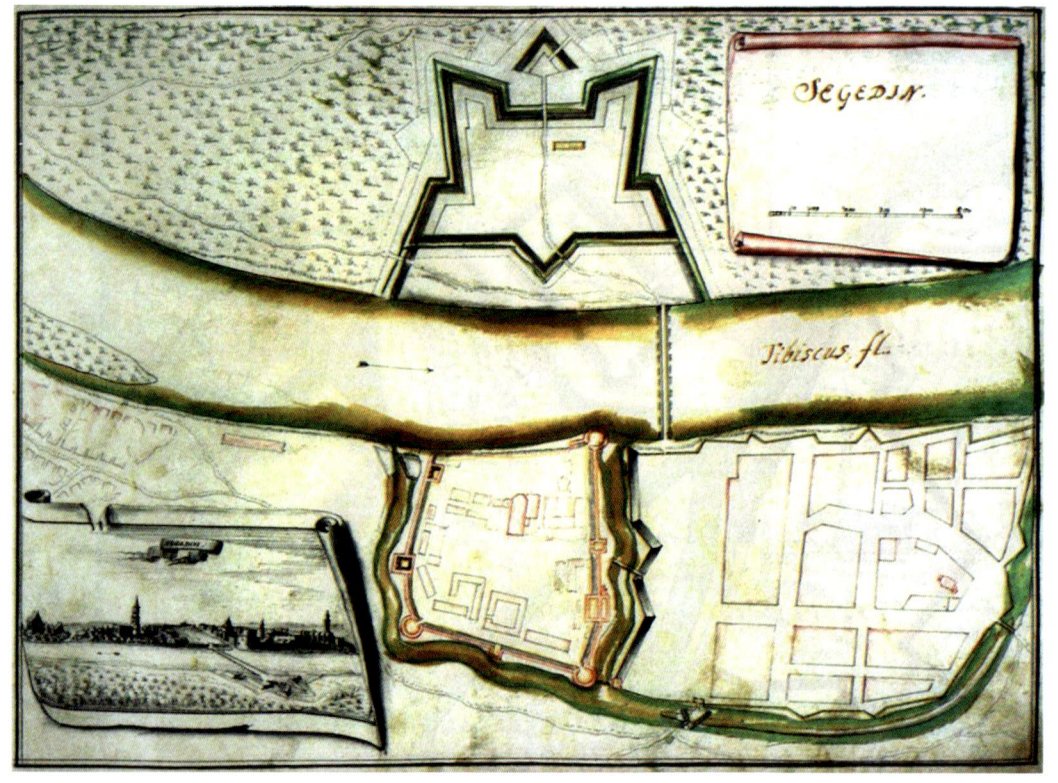

FIGURES 4.2A AND 4.2B. Fortress plans prepared by Ferdinando Marsili, depicting Segedin. Courtesy of the Bologna University Library (BUB Fernando Marsili, MS 21, nos. 28 and 40).

FIGURE 4.3. A general map prepared by Marsili based on his cartographic work illustrating Ottoman–Habsburg borders according to the Peace of Karlowitz, dated 1700. Courtesy of the Bologna University Library (BUB Fernando Marsili, MS 47).

uti possidetis, the future of fortresses—whether they would be retained or demolished—would be one of the main concerns in the negotiations. In addition to the citadels, Marsili illustrated usable routes, bridges, docks, and islands on the rivers.

Marsili prepared his maps to help the Austrians and the Venetians establish "secure" and "feasible" borders, as well as to "make the new Habsburg frontier impregnable before the peace conference got underway, by building a number of new citadels."[32] But these charts and maps were also designed to convince the Ottoman delegation of the Austrian and Venetian proposals. Accurate maps were key to achieving a comprehensive border delineation under the principle of *uti possidetis*. Once the parties agreed

on which places were under their control after the war, they could draw up new boundaries quickly. The surveying commissioners could then be sent to the region to mark the borders on the ground.

The maps that Marsili and his team prepared for the congress were based on yearslong surveys of the borderland. Marsili had studied topography and demography, rivers and mountains, fortresses and towns from the Adriatic to the Black Sea (figure 4.3). The Habsburg diplomatic establishment considered Marsili's cartographic program to be an accurate and authoritative representation of the region. But as they would soon learn, the Ottomans had ideas of their own.

Sulhnâme by Rami Mehmed Efendi

To understand how the Ottomans approached border demarcations, it is useful to turn to a text entitled *Peace Treatise* (*Sulhname* or *Vaka-ı Maslaha*), written by Rami Mehmed Efendi, the plenipotentiary of the Ottoman state during the Karlowitz negotiations. During the congress, Rami was the *reis efendi* (chief bureaucrat dealing with foreign correspondences), a member of the peace party in Istanbul. A successful career bureaucrat specializing in foreign and fiscal affairs, he would later be appointed the grand vizier. His tenure as grand vizier would end with the popular revolt in 1703, which was, in part, a reaction to the Karlowitz treaties.[33]

After the Karlowitz treaties were signed, the peace party was heavily criticized by various groups in Istanbul, not only because the Ottoman Empire ceded the provinces in Central Europe and Adriatic to its enemies, but also because the empire accepted a convention of European diplomacy, *uti possidetis*. Rami needed to defend the treaties signed in Karlowitz by showing that, in fact, the Ottomans did not blindly accept European norms, but rather thoroughly negotiated them. He intended to document that the Ottomans crafted the Karlowitz negotiations with their own visions of borders, sovereignty, and territory. Effectively, the *Peace Treatise/Sulhname* was written to illustrate how the Ottoman delegation refashioned the principle of *uti possidetis*.[34]

After a long prelude narrating how the negotiations started, Rami discusses why eternal and resilient peace (*sulh-ı ebed . . . mü'ekked sulh ve salah*) is preferable to temporary peace (*sulh-ı müddet*) and cease-fires (*ta'til-i sefer*).[35] Peace, he writes, should be based on mutual understanding, interest and trust. It should also uphold the security, well-being, and welfare (*'ibadullahın asayiş ve istirahati ve ma'murluğu*) of the people.[36] Enduring peace requires the cessation of frontier conflict and disorder, as well as a reduction of military fortifications in the borderlands.[37] To be sure, the principle of *uti possidetis* (*ala halihi zabtında olan zabtında kalsın*) should be accepted by the parties. But peace is a difficult task (*sulh ağır iştir*),[38] one that requires carefully considering the well-being of the people.

Throughout the text, Rami refuses to accept the principle of *uti possidetis* blindly. The Austrians, he notes, intended to implement *uti possidetis* without any qualification or exception (*istisna olmasın*).[39] For the Ottoman delegation, however, the principle had to be implemented in different ways in different cases, depending on the local economy, the nature of the border (whether it followed natural features or not), and security concerns.

The Methodological Disagreement

Rami's refusal of an all-applicable method in delimitation reflected a disagreement that surfaced between the Ottoman and Austrian missions at the very beginning of the conference. The Habsburg delegation proposed that the negotiations should start with verification of the postwar situation in accordance with the principle of *uti possidetis*. They declared that they would like to present a "paper" listing the places under their possession. If the Ottoman delegation agreed with its contents, major issues concerning borders would be solved at the conference with alacrity. If there were issues of disagreement, then the parties would negotiate these matters.[40]

The nature of the disagreement between the Ottoman and Austrian missions is spelled out in the protocols of the congress housed in the Haus-, Hof- und Staatsarchiv in Vienna. In the protocols, recorded by the Austrian delegation in Italian, we hear the voice not of Rami but of Alexander Mavrocordato, the Ottoman dragoman.[41] What follows is a translation of the dialogue from Italian between Count Leopold Schlick, the assistant to Count Wolfgang Öttingen, and Mavrocordato, dated November 13, 1698. Mavrocordato begins by articulating the Ottoman position.

MAVROCORDATO: We should leave the commissioners to make the artificial [boundaries] over the places where it is left to them to make the local denomination. In substance, they have the same intention, but your excellences want it in the opposite order.

COUNT SCHLICK: The distinction of his excellence is well understood[;] it is true that the delineation of borders must be the last [step] of the peace, but not already of the treaty. To remove the difficulty from your excellences, we are happy to leave the delineation of borders to the very end. But let's start with possession, determining how much and what is in the possession of one and the other party.

MAVROCORDATO: We agree that the possessions should be specified first. From there we will come to establishing the boundaries[,] and we will try to determine them as clearly and distinctly as possible.

COUNT SCHLICK: Given the readiness on the part of your excellences to agree, we want to express our possessions in writing [*in scritto*] to the gentlemen negotiators, so that your excellences can consider them and then make their feelings known.[42]

If we compare the protocols with Rami's *Sulhname*, we notice a difference. In Rami's text, we get the impression that the *paper* verifying possible new borders (i.e., which party held which region after the war) had already been prepared (*tahrir eylediğümüz sınır*).[43] In the protocols, however, Schlick states that they would like to put the possessions in writing, implying that there was no document yet illustrating borders in detail, or that they were in the process of preparing it. In any case, the Ottoman mission claimed the Austrians insisted on solving most of the boundary problems on paper at the congress, even if they agreed that the final effort for peace might be left to the demarcation committees on the ground.

What was this *kağıd*, or document *in scritto*, Rami and Schlick referred to? Was this paper a chart based on Marsili's maps and accounts? Unfortunately, neither Rami's text, nor the protocols, nor Marsili's papers in Bologna allow us to fully answer these questions. Still, in light of the available evidence, we can deduce that the Austrians had prepared (or intended to prepare) a chart, perhaps with maps, listing the places (fortresses and towns) under the possession of the Habsburgs, the Ottoman Empire, and other states to start the negotiations.

One term that Rami used offers us additional insight. The Ottoman delegation clearly refused to start the negotiation based on a prepared document. Rami wrote that the Ottoman delegation pondered the Austrian proposal. When Rami mentioned *kağıd*, he probably referred to several documents attached to one another in the form of a scroll.[44] Rami then wrote that the Austrians tried to frustrate the Ottoman delegation's willingness to solve the problem with a *karaltı* concerning the demolition and evacuation of fortresses.[45] Interestingly, in one of the manuscript copies of the *Sulhname*, the term *karaltı* is marked with vowel diacritics to ensure its correct reading, suggesting that the term held a peculiar or even technical significance.[46] *Karaltı*, which is the cognate of *kara*/"black" in Turkish, means "an image that is well discerned or perceived." *Karalamak*, the verb form, refers to darkening a paper by scribbling, sketchy writing, or calligraphic exercise. Might *karaltı* refer to a graphic representation, cartographic plan, or sketch map?

The Ottomans had not used maps for delimitation, either for internal or external borders, until the late eighteenth century. Still, cartographic or diagrammatic plans of fortresses and battlegrounds were not unknown in Ottoman circles.[47] In Marsili's own collection, there is an Ottoman plan of the fortress of Buda, dated circa 1684.[48] The generic term for such illustrations was *resm* (drawing) or its cognate *tersim* (image making), *suret* (image) or its cognate *tasvir* (depiction). Large-scale maps or sea charts, on the other hand, were known as *harti* or *karti*, Ottoman Turkish renderings of the German word *Karte*.[49] I have not found any example in which *karaltı* was used for cartographic illustrations. Still, it would be reasonable to think that Rami might have referred to some hazy cartographic sketches or charts.

Rami also wrote that with this proposal, Austrians planned to frustrate the Ottomans and make them to lose their hopes (*me'yus etmek*). Why would *karaltı* demoralize the Ottoman delegation? Would the Austrians plan to show their superiority in geographic knowledge with some obscure illustrations as a tactic to undermine the Ottomans? He added that the real reason behind the Austrian tactic was that the Austrian delegation intended to avoid face-to-face (*mukaleme*) negotiations.

In any case, it was clear for Rami that a document prepared by the Austrians would be against the interests of the Ottoman state. Rami wrote that if the Ottoman side had accepted such an "inappropriate" *paper*, the negotiation would have been carried out within the framework that the document stipulated. The Ottoman delegation refused:

> If there is any possibility to carry out verbal negotiation, it is not suitable to exchange documents. Such a method was not [previously] allowed. Peace negotiation is a hard task and cannot be carried out by pen and paper alone. Each item should be negotiated face to face separately. One cannot prepare a paper for every matter and negotiate it

over and over. This would be a huge waste of time. We came here to converse; if the settlement had been possible by writing, there would have been no need for such a tough meeting.[50]

In short, both Rami and Mavrocordato thought that starting the negotiations with *kağıd*, *karaltı*, or the document *in scitto*, would result in a *fait accompli*. Accepting a document prepared by the Austrians listing or illustrating the statuses of the fortresses as an introductory framework would cause the Ottomans to start the negotiations at a disadvantage. The Ottomans categorically refused to do that, pushing instead for face-to-face talks. The negotiations required continuous deliberation on each element of the borders, case by case, until agreement was reached.

The Ottoman Approach: Inductive Method

For their part, the Ottoman delegation proposed a three-step border-making process. First, the parties should negotiate the principles of the overall peace settlement, including how *uti possidetis* would apply. Second, the parties could agree on obvious and well-known natural borders, such as mountains and rivers. The third and most arduous step was to work out the details for problematic regions. Where there were no obvious natural boundaries, or where complexities were introduced for different reasons, this step should be left to multilateral commissions in conversation with local people. Rami wrote, "This duty could not be accomplished by those who negotiated the treaty at the table, since the local situation is not known to them. . . . The deed should be accomplished on the ground [*iş zeminde suret-pezir olub . . .*]."[51]

> In some of the regions, borders were created by God. If there were such apparent and well-known marks, [that] could be used for the demarcation, we can decide at the table. But most of the places, this was not the case and the borders should be delimited by the representatives who were appointed for border making [on the ground]. In such cases, [the established norm is that] they invite old, respected, impartial[,] knowledgeable people from the local communities of two sides of the border and [have] them carry out the specifics of the demarcation.[52]

In response to the Habsburgs, he wrote:

> Your method stipulates that the borders and boundaries should be first charted and represented at the meetings, as an overview. In this approach, the particularities are handled later. Our proposal is the opposite. We propose to start from the particulars [*cüzziyat*] and handle each one separately. At the end[,] we will come to the best conclusion.[53]

Rami here affirmed that the Ottoman method moved from the particulars to the general, whereas the Habsburg method from the general to the particulars. Mavrocordato agreed:

> We will go from one end of the borders to the other step by step and decide: in this way we will

determine possession. Where there is an obstacle, we will remove it by demolition, evacuation, etc., so that based on our agreement[,] any demand for restitution be excluded. So, we will go gradually, and we will not make universal disputes where they are nothing but details because it is necessary to come to this delineation of borders by *induction.* The universal is too difficult [to decide]. Let's start with a particular from whichever point we wish.[54]

After formulating the inductive method, Mavrocordato underscored the principle that in most cases the demarcation should be carried out on the ground with input from local people.

It is worth notice that in the past[,] treaties have always laid down the formula for determining the frontiers [*un certo metodo circa i limiti*] but details were always left to commissaries named by the two parties. It would be impossible to look at these details of the settlement in a congress where the plenipotentiaries either have no knowledge of the country or, if they do, might well prejudice the negotiation of the whole treaty by their excessively lengthy debate. The best and indeed obvious solution is to refer the details of the proposed "artificial" boundaries to mature, expert[,] and sensible men who—proceeding quite differently from people here at the conference—can erect landmarks on the spot and make the boundary clear and unmistakable. This is the old expedient of our predecessors, and we should pay attention to the precedent.[55]

In Rami's and Mavrocordato's formulation, borders—at least, those without natural features—should be handled one by one in verbal negotiation, in accordance with legal precedent and with the participation of local residents. This was, they said, an inductive (*induttione*) method, from *cüzziyat* to *külliyat* or from *particolari* to *universali*. I argue that this statement was not only a tactical or procedural matter.[56] It was also an epistemological challenge. True knowledge about borders could not be produced at a congress table via texts, charts, or maps, since local conditions could not be appreciated from a distance. Local and customary practice trumped universal norms.

Rami Mehmed Efendi proudly stated that in the end, the Habsburg delegation accepted this "reasonable" (*ma'kul*) proposal. In Rami's view, this procedural win was as a diplomatic coup for the Ottomans. Local commissions would be composed of Ottoman and Habsburg delegations along with local notables, and the details of border demarcation would be left to them.

The Timișoara Case

A particularly difficult place for border drawing was the Banat valley in Transylvania. Surrounded by the Tisza, Maros, Timiș, and Bega Rivers, Banat lay between Hungary and Ottoman Transylvania (figure 4.4). It was a prosperous region, thanks to rich water sources and swamps fostering rice production. The city of Timișoara housed most of the population of Banat, which had been an Ottoman province since

1552. In 1696, the Habsburg armies occupied the province and captured the city. Although the Ottomans soon recaptured the town, much of the Banat valley remained under Habsburg control. The Ottomans reinforced the fort, which became the main Ottoman stronghold north of the Danube.[57]

In a report dated April 1697, Marsili wrote:

> The fort of Timisöara itself should be captured, either by force or by cession, or it should be dissolved by collective agreement. If dissolved, either it should be made equal with Timişoara [it would become a city without fortress] and other less important places, without superiority, or it should remain under the direct subjection of only one empire or another. The abandoned country shall be allowed to decide whether to remain intact or to be divided between the two empires.[58]

Marsili recommended to the Habsburg authorities that they establish control in the entire region if possible, including the city. If not, the fort of Timişoara should be demolished since it would threaten Habsburg Hungary. The question was how to partition Banat if the city of Timişoara was left in Ottoman hands.

Here was a challenging case. While most of the Banat region was under Habsburg possession, the capital, Timişoara, was under Ottoman control: Where should the border line be drawn? As noted above, Marsili first advocated natural borders. But the situation was complicated since the Austrians controlled most of the fortresses. Therefore, the Austrians advised various "artificial" lines through the grassland that separated the Ottoman and the Habsburg territories. A small portion of land, immediately surrounding Timişoara, would be ceded to the Ottomans in accordance with *uti possidetis*. The rest of the province would belong to the Habsburgs.

The Ottomans pushed back. They insisted that some lands were naturally connected (*muttasıl*) and could not be separated. The province of Transylvania, which lay in the Ottoman Empire, had historically been integrated with the province of Banat and the two were not separable. More important, if the capital city (*kürsi-yi memleket*) of a province and its hinterland were divided, the communities would suffer. Most of the population of Banat lived in the city of Timişoara under Ottoman control. The communities in Timişoara needed access to both the rivers and the rural hinterland to survive; the province should not be split but should remain integrated (*bütün ayırtlanmaya muhtaçtır*).[59]

The Habsburg delegation initially rejected this reasoning. Arguing that they had occupied the province by force, they insisted that according to the principle of *uti possidetis*, it was not acceptable for the Ottomans to ask them to cede the occupied lands and fortresses. While the Ottomans held on to the capital city, they noted, the Habsburg armies had captured several other fortresses and towns.

In the end, the Habsburgs ceded the fertile part of the province, known as Şebeş, to the Ottomans. They still argued, however, for keeping the Tisza and Maros Rivers, which ran along two sides of the region. They also insisted that there be no agricultural activity around the boundaries. The Ottomans pushed back once again, countering that the main aim of the peace was to restore the well-being and prosperity of the people (*ma'muruiyet*). Peace and prosperity should prevail together—but this could be achieved only if the land and the rivers were freely exploited by the people. The province of Banat was a natural entity; its borders had been drawn by God (*Temeşvar eyaleti lütf-ı hakla kendünden sınur bulmuş bir memleketdir*).[60]

In effect, the Ottoman delegation proposed a different interpretation of the principle *uti possidetis*. The principle should not violate natural dependencies of the populated city, nor should it trump the demographic, economic, and ecological unity of a province. Rami wrote:

> We told them[,] "The aim of our State is the rejuvenation of the cities, villages, and the countryside. However, [if we divide the province as you proposed,] the communities would not benefit from water." Then we asked them whether people should not drink water. The Austrians said they cannot prevent humans and animals from drinking water. Then we asked them: "So, if they live nearby rivers, should not they eat fish then?" "No," they said, "we cannot prevent them from fishing with boats, as hunting wild animals is blessed by God." Then we asked "Can fish be eaten without bread?"[61]

Through this step-by-step argument, the Ottoman delegation justified claiming the docks on the rivers in Banat province. Wheat was transported from farms by boat to the mills, which were located along the rivers. Since the closest river (the Timiş) was insufficient to provision the city, the borders should be drawn to include the more distant Tisza River as well.

Eventually, the Austrians accepted the Ottoman claim for Timişoara's hinterland. In return, the Ottoman delegation withdrew their earlier claims that Transylvania was inseparable. As formulated in article 2 of the Karlowitz Treaty, the parties agreed that the Ottoman possession would extend to the Tisza and Maros Rivers. The fortress of Timişoara was retained, while the Habsburg forts along the Tisza and the Maros were demolished. The communities on both sides would have the right to use the Tisza and the Maros for fishing, the watering of cattle, and other activities necessary for their livelihood. Boats would have free passage on the Tisza and Maros. Both Ottoman and Habsburg subjects could establish mills along the rivers. The islands in the rivers would remain in the possession of the Austrian emperor.[62]

Timişoara was a showcase for the Ottoman premise that particulars could not be sacrificed to general norms. *Uti possidetis* was feasible only if the local and natural conditions were taken into consideration. Most strikingly, the Ottomans made a moral-economic and even moral-ecological argument. They refuted the implementation of the principle *uti possidetis* categorically if the result would be harmful for people and averse to the natural integrity of the region. Isolating the city of Timişoara from its natural and historical hinterland was simply untenable.

The Congress of Karlowitz continued for two and a half more months. The Ottoman–Venetian, Ottoman–Polish, and Ottoman–Russian borders, along with such vexed matters as the status of merchants and captives, required extended discussions. The treaties were finally signed between January 1699 and July 1700. At that point, the borderlands commissions moved to their respective regions for delimitation from the Adriatic coasts to the Ottoman–Polish border. The delimitation process took two years. During this time, the surveyors visited the border towns, summoned local notables, and asked for details about local conditions, rivers, grasslands, and swamps. They erected stones if there was no natural mark. The European cartographers in the commissions prepared the maps, while the Ottoman scribes described the borders in detail.[63]

Conclusion

This chapter has focused on two competing negotiation strategies and understandings of border delimitation in the Karlowitz Congress at the turn of the seventeenth century. The Austrian–Venetian mission proposed to move from the universal to the particular, from representations (maps) to the represented (borderland localities). They claimed that only with this method could disputed areas be handled properly and settled efficiently. The Ottomans, by contrast, proposed what they called an *inductive* method. First, the local realities should be established on the ground; only then could maps, charts, or texts be configured as authoritative documents. The Ottoman side declared that this method alone ensured the prosperity of the borderland communities—and the prosperity of its people, after all, was the ultimate object of the state.

How should we understand these contrasting arguments? As I have already suggested, the Kar-

lowitz moment took place not only during the cartographic revolution in Europe but also during the consolidation of territorial states with new notions of borderlands.[64] Therefore, the debate between the Ottoman and Austrian–Venetian missions can provide us with insights about how different actors crafted arguments concerning state-making, territoriality, and borders at this early stage of the political-cartographic turn. This is why the contrast between the deductive method of the Austrian–Venetian diplomats and the inductive methods of the Ottomans is worthy of study.

The Ottoman refutations of deductive, top-down, abstraction-oriented border negotiation in favor of inductive, bottom-up, local-oriented border making was not simply a tactical move. The Ottoman position was based on an epistemological and moral critique of the cartographic-political technology of the Austrian–Venetian side. It was an epistemological critique to the extent that the Ottomans made a case against the idea that diplomatic documents, charts, and maps could accurately represent the local conditions. Accuracy lay in local knowledge and experience, not in abstraction. At the same time, theirs was a moral critique, inasmuch as a deductive method would damage the well-being of the communities. It would not be an overstatement to argue that the Ottomans crafted a moral-ecological case for constraining the use of maps; *a priori* representations must be resisted, lest they undermine the complex resource webs that sustain life in the borderland.

PART II

Pushing Back against the Sovereign Map

5

Alexander B. Murphy and Cy Abbott

Territorial Challenges at Interstate Borders
Where and How History Matters

In theory, the conventional world-political map depicts a planet of clearly defined geopolitical spaces, each one separated from the other by a distinct border. Yet there is an enormous disconnect between that map and actual or aspirational spaces of sovereign territorial control. That disconnect, which the conventional map completely obscures, finds expression in myriad territorial disputes challenging the existence or fairness of international borders—whether land or maritime.[1]

The stakes of this disconnect are high. As a variety of studies have confirmed, territorial disputes between states are more likely to lead to serious disagreements and even war than any other source of tension.[2] Unsurprisingly, then, interstate territorial conflict has become a major focus of study in recent decades, with scholars from multiple disciplines seeking to understand relevant catalysts and consequences.

Some time ago, one of the authors of this study sought to contribute to that effort in the form of a 1990 study exploring how dominant ideological and legal norms influence the geography of interstate territorial conflicts. The core argument advanced in that study was that only one form of justification for a ter-

ritorial claim is clearly in keeping with those norms: a historical argument that "the disputed territory is rightfully 'ours,' . . . it was illegally taken away from 'us' and 'we' have the right to reclaim it." The study goes on to argue that the primacy of that form of justification greatly influences which territories are in dispute, the spatial imaginaries that shape thinking about possible conflict outcomes, and the global geographical distribution of territorial conflict.[3]

Developments over the past three decades have confirmed the basic thesis advanced in that study, but the recent settlement of a few seemingly entrenched territorial conflicts raises an important question that goes beyond the prior study's scope: Why are some historically justified territorial claims more amenable to settlement than others? Addressing that question requires comparative thinking about settled and not-settled cases. Contextual differences greatly complicate such an effort; however, our efforts in this vein led us to focus on a contextual variable that is often acknowledged but rarely given great credence: the role played by differences in the ideological foundations of dominant territorial imaginaries.

Our central claim in this chapter is that territorial conflict resolution is particularly difficult in the face of territorial imaginaries rooted in long-standing, deep-seated ethno-nationalist conceptions. Such conceptions resonate with a core principle that emerged along with the modern state system: namely, the notion (encapsulated in the nation-state ideal) that the state should embody a distinct ethno-national community. We suggest that those seeking insight into the territorial dynamics that the conventional political map successfully conceals could productively pay more attention to the role played by the kind of foundational, enduring, ethno-national territorial imaginaries that are demonstrably at play in many of the world's most intractable territorial conflicts today (Serbia's claim to Kosovo, Russia's to part or all of Ukraine, China's to Taiwan, etc.).

To make the case, we begin by examining why historical ethno-national territorial imaginaries are so powerful and resistant to change. We proceed to ground the discussion in a contrasting analysis of two deeply contentious twentieth-century disputes over territory: that between Greece and Turkey and that between Ecuador and Peru. Curiously, the Ecuador–Peru case is the one associated with more overt violence over the past hundred years, yet it is the Greece–Turkey dispute that has been the more resistant of the two to settlement. Explaining this conundrum is complicated by myriad contextual differences that distinguish the two cases, but one factor is particularly helpful in making sense of it: the profound differences in the territorial imaginaries at play in each dispute. In the closing section of the chapter, we argue that the powerful influence of imaginative differences in these two cases points to the need for a greater emphasis more generally on the symbolic dimension of territory in our studies as well as in our cartographic renderings of sovereignty.

The Importance of Historical Ethno-nationalist Territorial Imaginaries

Much of the literature on territorial conflict places primary emphasis on such factors as the physical properties of disputed territories, the relations among the disputants, the types of legal or governmental systems that exist in the disputant states, and the role international institutions can play. Tellingly, Emilia Powell and Krista Wiegand also note the potential influence of territory's "symbolic, nationalist value,"[4] though the modest attention this issue receives in their literature review suggests it is not a central concern of most studies of territorial conflict abatement. Instead, the literature on territorial conflict tends to emphasize the importance of tangible-geographic, governmental, and institutional circumstances.

The tangible-geographic circumstances receiving sustained attention include the distribution of economically valuable resources, population characteristics, and the strategic properties of places and spaces.[5] Much of the literature on governmental regimes focuses on the propensity for authoritarian, as opposed to democratic, regimes to go to war over territory.[6] Institutional studies tend to emphasize such matters as the role played by formalized international alliances,[7] the influence of different legal arrangements,[8] and the impact of domestic institutions on the potential for territorial conflict.[9]

For all the insights such studies offer into the catalysts and causes of many interstate territorial conflict, mixed findings are not uncommon (notably, on the role played by regime type). As such, some commentators point to the need to look beyond material and institutional factors.[10] That means focusing on territory's intangible, symbolic dimensions. Yet most efforts in this regard are limited in scope and analytical depth. Studies that consider the importance of ethnic ties represent a partial exception.[11] Yet the emphasis in such studies is mostly on concrete circumstances (e.g., the influence of ethnic distributions on territorial ideology)—leading Harris Mylonas and Maya Tudor to conclude that "research illuminating the comparative modalities of nationalism has been historically sparse."[12]

Taking the symbolic dimensions of territory seriously requires consideration of the different foundations on which territorial imaginaries are built. In some places, they are grounded in the right of a historically imagined ethnocultural group to have a political space—a nation-state—of its own (e.g., France, irrespective of the fact that there have always been many people in France who do not identify as French). In other places, they have unavoidable roots in colonial territorial divisions, meaning that the "nation" part of the nation-state dyad is cast more in civic than in ethnocultural terms (e.g., Nigeria). Paying attention to such differences represents a potentially important way of deepening the engagement with nationalism as a factor affecting territorial disputes.[13] Consider, for example, the contrast between states such as Spain and Hungary, where the idea of

the state was closely associated with a conception of the nation in the original sense of the term (a group of people with a shared sense of history, culture, and ethnicity), and Gabon, where the concept of the nation could only be thought of in civic terms.[14]

The point is that some states are less positioned to advance territorial imaginaries rooted in a deep-seated conception of an ethno-historical nation than others. As such, they are not in a strong position to tap into the powerful ways in which cultural-historical imaginaries shape a sense of identity or place.[15] Territorial imaginaries in such states necessarily have a more instrumentalist character—grounded, for example, in the configuration of preexisting administrative units or in physical-environmentalist conceptions of areas that are widely viewed as discrete spaces (an island, a peninsula, etc.).

To be sure, states in this position can, and often do, seek to promote a nationalist discourse based on the distinctive characteristics of a polity's population, and that discourse often helps foster a sense of ethnic particularity (Chile is a country for Chileans, Argentina for Argentinians, etc.). But the starting point in these cases is a territorial construct that lacks the deep-seated historical-ethnocultural referent found in circumstances where states carve out spaces on the political map based on specific territorial claims made on behalf of distinctive (if imaginative) ethno-historical groups. Such claims are typically rooted in founding stories premised on the ethno-historical distinctiveness of those groups. A comparison between the territorial struggles that have taken place between Greece and Turkey and between Ecuador

FIGURE 5.1. The Megali Idea, as represented in a widely circulated cartographic image in Greece in the 1920s. Published and distributed by Pelasgos Publications, Athens, Greece. Public domain.

and Peru suggests that ethno-historical formulations of the nation tend to make territorial imaginaries less open to negotiation or modification than do more instrumentalist conceptions.

The Greece–Turkey Case

The Greco–Turkish territorial conflict is a product of national and territorial imaginaries with historical ethnocultural-cum-territorial roots that have served to keep the conflict alive—even in the face of a formal geopolitical alliance between the two countries and numerous attempts at third-party mediation.[16] Before modern Greece and Turkey came into being, many Eastern Orthodox speakers of Greek dialects and Turkish-speaking Muslims lived in close proximity to one another, and identities and associated territorial conceptions were somewhat flexible. However, with the rise of the modern Greek and Turkish states in 1830 and 1923, respectively, national identities took on a fixed, spatially exclusive character rooted in a confluence of deep-historical and modern conceptions of nation and territory.

The cultural-historical roots of Greece's and Turkey's contested territorial imaginaries can be traced back to claims each state makes of succession to the Roman Empire, with Greek conceptions derived from the Eastern Roman (Byzantine) Empire and Turkish conceptions from the later Ottoman Empire.[17] Over the course of their existences, these empires occupied almost entirely coextensive territories in the Eastern Mediterranean. Both the Byzantine and Ottoman Empires nurtured the Roman legacy—

making it possible centuries later for that legacy to form a part of the ethno-territorial imaginaries that evolved along with the modern Turkish and Greek states.

Modern Greece and Turkey emerged out of the explicitly nonspatial sovereignty system of the Ottoman millets.[18] The millet system divided non-Muslim people into categories based on religion but gave each group the ability to have its own courts and live according to its own legal rules. The largest millet, the Millet-i Rum (Roman millet), encompassed many non-Muslim people, including Greeks who maintained separate social institutions and had distinct legal rights, all the while living among Muslims speaking varied Turkish dialects. That was certainly the case in the Ottoman Empire's port cities, where Western European conceptions of nations and nation-states began to animate Greek populations in the Ottoman Empire. Those cities became the hearth, if not the stronghold, of revolutionary thought that sparked the War of Greek Independence (1821–1829), which gave birth to the modern Greek state.

From its founding, the Greek state was shaped by the nation-state ideal, meaning that prevailing territorial imaginaries were grounded in a notion of a Greek people speaking a distinctive language, following a particular religion, and sharing a common history.[19] To be sure, the imagined political-cultural progenitor of the Greek state was internally disputed. Some championed "Byzantinism"—viewing the modern state as the successor state to the Eastern Roman Empire. Others argued for "Hellenism," the more revolutionary and Enlightenment-influenced

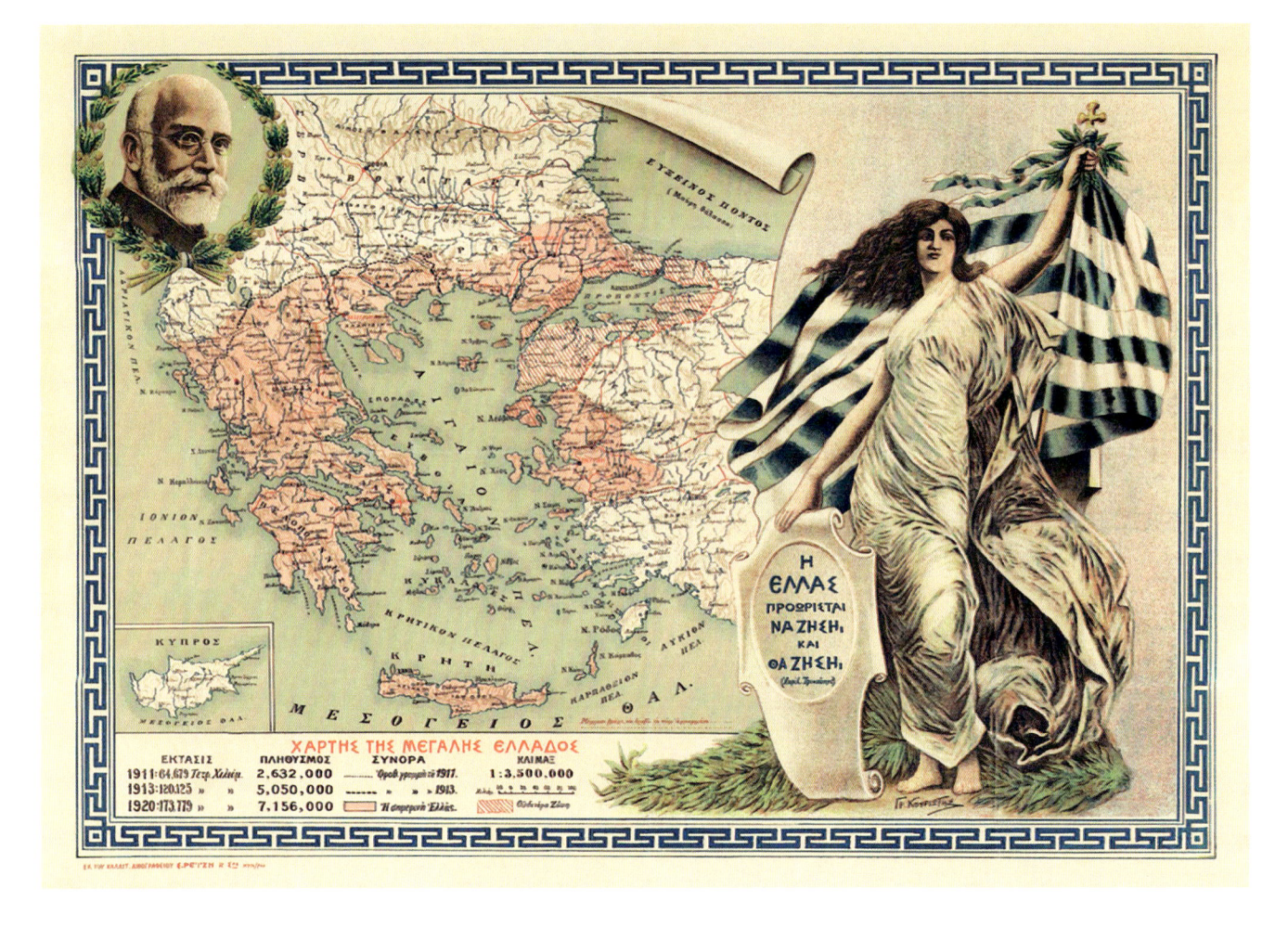

idea of modern Greeks being heirs to the culture and territory of the ancient Greeks, with associated claims to having given birth to Western Civilization.[20] Despite their differences, however, both conceptions were grounded in a historical-ethnocultural territorial imaginary that encompassed places and people lying well beyond the Peloponnese peninsula, Attica, and the Cyclades. Indeed, until 1922, there were more Greek inhabitants of the Aegean port city of Smyrna (İzmir) along the Turkish coast than there were inhabitants of the Greek capital of Athens. Irredenta thus became a foundational territorial aspiration of modern Greece—leading to the embrace of the Megali Idea. The power of that idea was reflected in nationalist cartography (figure 5.1), and it fueled an expansionist project that saw Greece enlarge its borders at the expense of the weakening Ottoman Empire between 1881 and 1913.

The stage was further set for the modern Greco–Turkish territorial conflict by the collapse of the Ottoman Empire, the 1919–1922 Greco–Turkish War, and a subsequent massive population exchange in 1923. The late Ottoman Empire found itself racked by intercommunal violence and instability brought about by the growth of nationalist movements throughout the Balkans. The development of modern Turkish nationalism and the formation of the modern Turkish Republic were the products of dramatic territorial contraction.[21]

Under the leadership of Mustafa Kemal, who

would later be bestowed with the honorific Atatürk (father of the Turks), a modern and robust Turkish nationalism was established that saw Turkey stave off large-scale territorial dismemberment at the hands of the victorious Allied forces of the First World War. The consolidation of the modern Turkish state represented a clear challenge to Greece's Megali Idea, which encompassed not only both sides of the Aegean Sea but extended to Constantinople (İstanbul). With Greece and Turkey's territorial imaginaries in direct opposition to each other, the previously flexible identities of Greek and Turk became increasingly divergent and fixed. That fixity found concrete expression in the pursuit of ethno-national irredentism by the Greeks and by efforts to exert control over as much of the core Ottoman territory as possible by the Turks.

The ensuing conflict resulted in extraordinary loss of life and mass dispossessions.[22] In the face of this human tragedy and with no clear path forward, Greece and Turkey signed the Treaty of Lausanne in 1923, which called for the Eastern Orthodox inhabitants remaining in Turkey to be classified as Greeks (regardless of linguistic or cultural affiliation) and forcibly "repatriated" to Greece, and for any Muslims remaining in Greece to be classified as Turks and to be deported to Anatolia.[23] The resulting reshuffling of individuals and communities produced a cultural-geographic arrangement on the ground that conformed more neatly to the developing territorial imaginaries of Turkey and Greece—infused as they were by the nation-state ideal.

The land border between the Republic of Turkey and the Kingdom of Greece[24] was delineated as the course of the Evros (Meriç) River in Thrace, running from the Bulgarian border to its mouth in the Aegean, with a small strip on the western bank opposite the city of Edirne (Karaağaç Triangle) allotted to Turkey. The maritime boundaries of the states were also codified, with Greece abandoning all claims on Anatolia but retaining sovereignty over the numerous Aegean Islands to the west—with the exception of Imbros (Gökçeada) and Tenedos (Bozcaada) flanking the entrance to the Dardanelles, and the then-Italian-controlled Dodecanese, which would later be turned over to Greece after the Second World War.

That arrangement did not settle the Greco–Turkish territorial dispute, however. Instead, competing sovereignty claims have skirted around the constraints of the treaty, leading to a shift in the locus of contestation—mostly from the terrestrial to the maritime arenas, but also involving a few small islands and having proxy components. One notable exception is the land border along a portion of the Evros/Meriç River, which has been a flashpoint of violence, with several small-scale deadly clashes having taken place in recent years between Greek and Turkish border guards.[25]

The most significant clashing conceptions of territorial sovereignty, however, have focused on resource and sovereignty rights to the waters and airspace that lie between Greece and Turkey;[26] those conflicts even gave rise to a recent, short-lived movement to claim a Turkish "Blue Homeland" (Mavi Vatan). That term, coined by Turkish naval staff officers Cem Gürdeniz and Cihat Yaycı in 2006, refers to

FIGURE 5.2. A September 2019 press conference given by Turkish President Recip Tayyip Erdoğan featuring a map of the "Blue Homeland." *Aydınlık*, Istanbul, Turkey. Public domain.

a more expansive spatial conception of Turkey's maritime rights than had been publicly advanced for decades. Late in 2019, the term came to play a somewhat conspicuous role in official Turkish political circles, fueled by cartographic depictions that were publicly showcased by President Recip Tayyip Erdoğan and that were used to support Turkey's claim to an expanded exclusive economic zone before the United Nations (figure 5.2). Not surprisingly, this posturing by the ruling Justice and Development Party was met with vehement opposition in Greece and was widely dismissed by the international community as a gross violation of existing sovereignty norms—leading to a recent turn away from the idea in mainstream Turkish political circles. But the emergence of the Blue Homeland idea, and Greek reactions thereto, speak to the continued volatility of the Turkey–Greece territorial dispute.

The point is that it is all but impossible to make sense of the dispute's inertia without giving due attention to the mutually exclusive, oppositionally constructed territorial imaginaries of Greece and Turkey—grounded as they are in deep-seated historical, ethnocultural-cum-nationalist narratives. The locus of the conflict has shifted over the past century, but the underlying cultural and territorial sensibilities that sustain it have not.

The Ecuador–Peru Case

Over the past two centuries, the Ecuador–Peru territorial dispute stands out as one of the most serious modern conflicts over territory in the Western Hemisphere. At least thirty-four bloody border clashes have broken out between the two countries since Ecuador's secession from Gran Colombia in 1830,[27] including serious fighting in 1941, 1981, and 1995. Despite the conflict's duration and intensity, Ecuador and Peru entered into an agreement in 1998 that effectively brought hostilities to an end. The agreement sparked protests in some quarters and sporadic localized tensions remain; but the territorial dispute is no longer a major flashpoint between the two countries.

Why has it been easier to settle this serious, entrenched conflict than the Greece–Turkey territorial dispute? There can be no easy answer to that question, but one factor of notable—potentially even overriding—significance is the different nature of the identity-nationalism-territory relationship that exists in the two cases. To be sure, nationalist passions are and have been strong in both cases. However, nationalist identity and associated territorial imaginaries in the Ecuador–Peru case emerged from circumstances that were not conducive to the development of a territorial imaginary grounded primarily in a deep sense of historical-ethnocultural distinctiveness, as was the case in Greece and Turkey.

This is not to suggest that Ecuador and Peru are immune from the cultural and territorial ideals that developed along with the modern state system.[28] In both countries, nation-building initiatives have fostered strong senses of national identity among at least a significant portion of their constituent populations.[29] Moreover, conflicts between the two states have served to stoke nationalist feelings and make territory a central component of national identity.[30] But in the face of a historical-geographic backdrop that was not conducive to the development of an ethnoculturally grounded, nationalist-cum-territorial imaginary, nation building in Ecuador and Peru necessarily came to be tethered to political-geographic spaces carved out by an external political power—two spaces characterized by a mix of Indigenous, Mestizo, and largely European people with no clear ethnocultural divide separating the two.

To be sure, Peru could imbue its nation-building story with an ethnocultural veneer—casting itself as the modern polity of the Incan people. But with more than 60% of its population a mix of White and varied Amerindian peoples, and another 25% of its citizens made up of Indigenous peoples, many of whose ancestors were conquered by the Incans in the past,[31] Peru has never been in a strong position to construct a clearly defined nation-building discourse cast primarily in historical-ethnocultural terms. As for Ecuador, its even higher percentage of Mestizos and smaller Amerindian population make such a case even harder to sustain. And there is no clear basis for

distinguishing the two countries on ethno-linguistic or ethno-religious grounds; in both Ecuador and Peru, Spanish is the dominant language, Catholicism the dominant religion, and Mestizo the dominant ethnicity (if it can be thought of as such).

Against this backdrop, nation building has been rooted in what Anssi Paasi calls state-driven "spatial socialization"[32]—a process that was less about advancing grand ethnocultural claims than about promoting the interests and integration of preexisting territorial constructs through the creation of national-scale social and institutional projects, the development of infrastructure, the establishment of national education programs with offerings on Ecuadorian and Peruvian history and geography, the dissemination of nationalist iconography, and the growth of an increasingly powerful public sector that seeped into peoples' lives.[33] Part and parcel of spatial socialization was the development of territorial imaginaries in Ecuador and Peru that helped fuel nationalism because they were in tension with each other.

The details of how that tension played out over time are beyond the scope of this paper.[34] To summarize, under Spanish colonial administration, most of South America was initially assigned to the so-called Viceroyalty of Peru. Over time, the challenge of administering such a large territory as a single unit became so great that the Spanish divided it into several *real audiencias* in the mid-sixteenth century. In the west-northwest part of the continent, one of these was centered on Lima and another on Quito (the first political-geographic arrangement from which the modern states of Peru and Ecuador can trace their

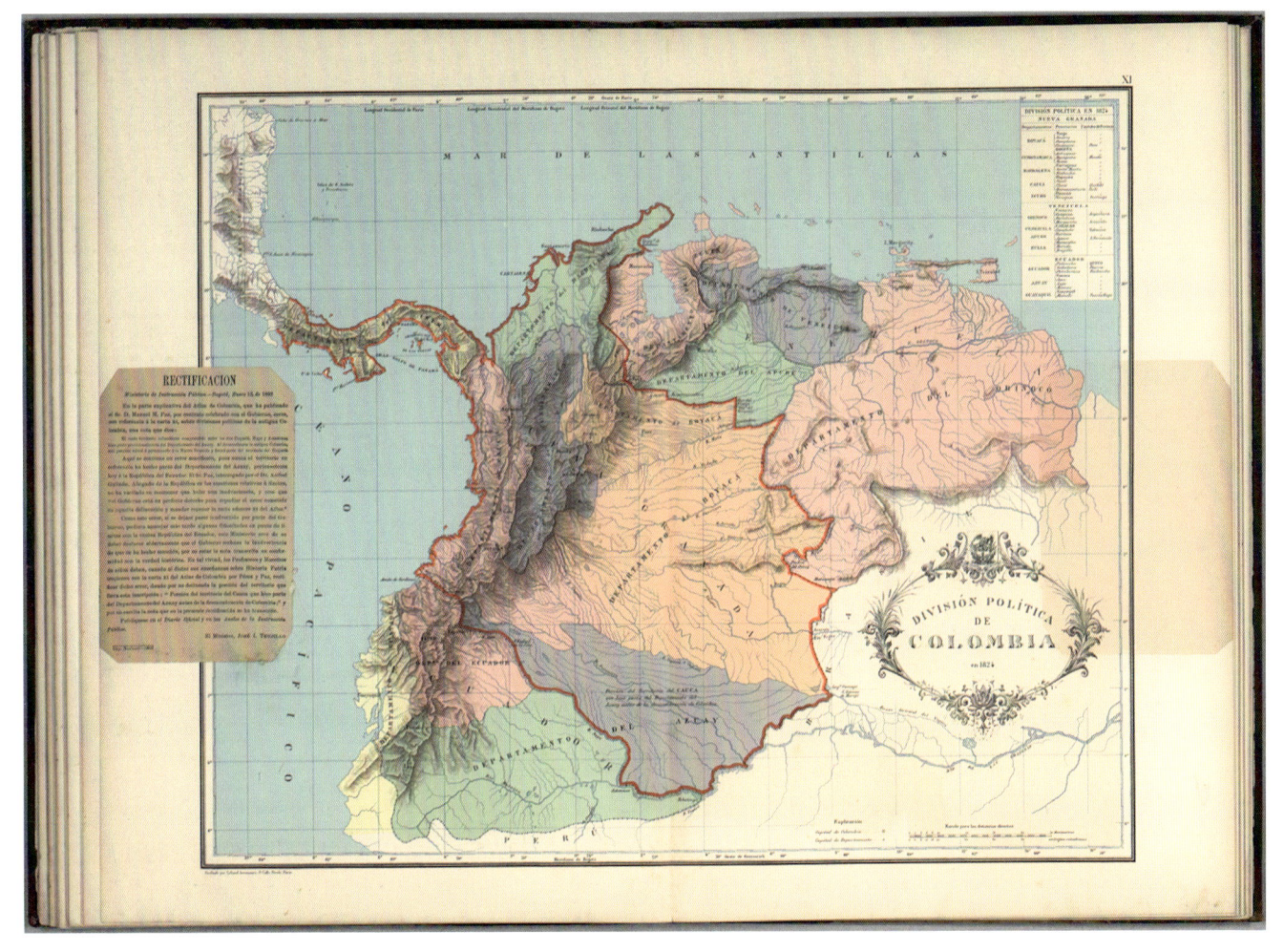

FIGURE 5.3. A Colombian-commissioned map superimposing the republics' 1889 boundaries atop the former Viceroyalty of New Grenada. The former *Departmento del Azuay* in the Amazonian interior was split between Colombia and Ecuador in the dissolution of Gran Colombia. These two countries accepted the split, but Peru's claim to, and then occupation of, the *departmento*'s eastern portion set the stage for the Ecuador–Peru territorial conflict. A. Lahore, *Division politica de Colombia en 1824*, 1889. Courtesy of the David Rumsey Historical Map Collection, Stanford University. Available at https://purl.stanford.edu/hy310dw6362.

lineages). However, in the early eighteenth century, the Spanish decided to incorporate the Quito *audiencia*, together with remote, unmapped territory in what is now northwest Peru, into an overarching Viceroyalty of New Granada that extended into present-day Panama.

What followed is a complex history of territorial modifications, including a short-lived 1723 decree that returned control of what is now northwest Peru to the neighboring Viceroyalty of Peru for sixteen years and a number of other decrees modifying the original demarcation of New Granada and establishing ecclesiastical territories with varying degrees of actual authority in the region. Then, in 1819, in the wake of independence movements sweeping South and Central America, most of what had been the Viceroyalty of New Granada emerged as the independent Republic of Colombia (figure 5.3).

FIGURE 5.4. The territory in dispute between Ecuador and Peru during the nineteenth and twentieth centuries. The map shows the small area over which Ecuador gained effective control in the wake of the 1998 settlement of the dispute. Original map created by the authors.

The border between Gran Colombia and the newly independent state of Peru left what became Ecuador, as well as the territory it claimed in northwest Peru, in Gran Colombia. Hence, when Ecuador separated from Gran Colombia, it claimed the right to all the territory shown in figure 5.4—a claim it viewed as having been formally ratified in an agreement signed at the time of independence. Peru, however, contended that a Spanish decree of 1802 transferred control over the disputed territory to the Viceroyalty of Peru, and Peru had the power to effectuate that claim—bringing the territory in dispute under Lima's *de facto* control.

As Isaiah Bowman put it, "The sum of the matter, so far as legal rights are concerned, is that each side can make out a plausible and to some degree a juridical case for its claims."[35] That set the stage for a series of confrontations between the countries that were temporarily suspended when the 1942 Rio Protocol was signed in the aftermath of serious clashes during which Peru gained the upper hand. Brokered by outside powers (the United States, Brazil, Argentina, and Chile), the protocol left Peru in control of most of the territory in dispute, but required Peruvian forces to withdraw from Ecuadorian territory they had invaded during the preceding conflict, gave Ecuador free navigation access to the Amazon basin, and left to later resolution a few disputed places along the border.

It was not long, however, before Ecuador began contesting the Rio Protocol—arguing that it was imposed under duress, that Peru was impeding Ecuador's navigation access to the Amazon, and that "new geographical information" regarding the extent of an important borderland watershed meant that the Rio Protocol border decision was based on spurious assumptions about underlying physical-geographic circumstances.[36] By 1960, Ecuador declared the protocol to be null and void—paving the way for more border clashes that culminated in a small-scale war in 1995 (the so-called Cenepa War). Neither side was able to gain much territory during the conflict, and the stalemate led to peace negotiations involving the same external powers that played a role in the protocol. These culminated in a 1998 agreement (the Brasilia Peace Agreement) that led to a reaffirmation of most of the Rio Protocol border while giving Ecuador effective control, though not formal sovereignty, over a small disputed area along the border (the Tiwinza region—see figure 5.4), recognizing Ecuador's claim to another small unmapped area along the border, granting Ecuador a few hundred acres of land along an Amazon tributary to build port facilities, and guaranteeing Ecuador's navigation and trading access to the Amazon watershed.[37]

A convergence of circumstances helped pave the way for this settlement: war weariness, the face saving for Ecuador that accompanied its ability to stand up to superior Peruvian forces in the Cenepa War, political instability, economic challenges that both countries needed to address, and intensive pressure along with strong-arm mediation tactics from the involved external powers.[38] None of these would have been possible, however, without a nationalist territorial imaginary that made a compromise possible— particularly on the part of Ecuador, which effectively ceded almost all its territorial claim. That such flexibility existed is in many ways surprising, given the long-standing, signal role the dispute with Peru had played in stoking nationalist sentiment in Ecuador. But the absence of a deep-seated ethnocultural foundation to Ecuador's territorial claim arguably made a

dramatic compromise possible when other circumstances were favorably aligned.

The foregoing contention is difficult to prove conclusively, but its plausibility is hard to deny when we consider nationalism's fundamentally cultural-ideological roots.[39] That plausibility is further suggested by myriad studies pointing to the seminal role ethnocultural sensibilities play in the construction of national identity[40] and by studies showing that emotional issues make territorial compromises more difficult to achieve.[41] Ethnocultural sensibilities are clearly one of those emotional issues, as confirmed recently by the outsize role ethnocultural referents play in contemporary right-wing populist-nationalist movements (e.g., the emphasis such movements place on perceived threats to traditional culture coming from the arrival of immigrants and from school teachings and books discussing racial oppression and differences in sexual orientation).

In the absence of any plausible way for Ecuador to position its nation-building case—and, by extension, its cultivation of a territorial imaginary—on ethnic or racial grounds, the rallying point for nationalist identity was the specialness of people living in a territory that had a distinctive past and a potentially glorious future. That gave rise to a nation-building project grounded in physical-geographic spaces: a coastal plain centered on Guayaquil (the first part of Ecuador to achieve independence), the ranges and basins of the Andes to the east of the coastal plain, and the so-called Oriente—the edges of the Amazon basin to the east that have long been controlled by Peru.

FIGURE 5.5. 5-centavo postage stamp ("Social and Rural Worker's Insurance Fund"), 1938. Issued by the government of Ecuador, this commemorative stamp shows Ecuador extending into Amazonia. Courtesy of the Smithsonian National Postal Museum. Available at https://postalmuseum.si.edu/object/npm_1977.1119.6.14.1.

In two articles examining Ecuador's "imaginative geographies," Sarah Radcliffe illustrates the importance of this tripartite regional vision to the country's nationalist-territorial project.[42] As she makes clear, the Oriente loomed particularly large—both because of its disputed status and because of its potential economic and strategic importance. Ecuadorian maps, writings, and other types of images regularly presented the Oriente as part of Ecuador; slogans such as "Ecuador Pais Amazonico" (Ecuador an Amazonian country) were painted on the sides of buildings; widely circulated geography and history books on the country treated the Oriente as part of Ecuador; and many stamps were adorned with maps showing Ecuador extending into the Amazon basin (figure 5.5). The result, following Radcliffe, is that for many Ecuadorians, "the issues of Amazonian access, territorial sovereignty and prospects for material development . . . were crucial ways through which they thought about themselves as Ecuadorians."[43]

For all the power of this territorial imaginary, its roots lay in inherited historical political-geographic arrangements, in interpretations of the legality and effect of Spanish colonial decrees, in resentment over Peru's seizure of the region, and in economic and strategic considerations. That does not mean that national identity in Ecuador lacks any cultural underpinning; as a fundamentally cultural-ideological force, nationalism has an unavoidable cultural dimension that, in the Ecuadorian case, is evident in the promotion of an official discourse about the country's people highlighting "the mutability of ethnic subject identities, and the supposed ease of entry of different groups into a homogeneous *Mestizo* identity, with a strong dash of indigenous folklore."[44] In contrast to the deep-seated, emotionally laden, historical-ethnocultural ideologies that go to the heart of the Greece–Turkey territorial struggle, however, the cultural elements of Ecuadorian nationalism cannot and do not articulate strongly with a deep-seated historical-ethnocultural story. The consequence is that, under the right circumstances and in the face of the right pressures, a settlement between Ecuador and Peru could be worked out—one that, though not accepted by all, has endured for more than two decades.

Discussion and Conclusion

Nationalism's role in sustaining territorial conflict is well established, but not all forms of nationalism have the same impact. Instead, territorial imaginaries based on historical-ethnocultural distinctiveness have psychological and emotional impacts that can make conflict settlement especially difficult. Those sensibilities are clearly at play in most of today's serious, persistent territorial disputes: Japan and Russia, Japan and Korea, China and several neighboring states, India and Pakistan, Armenia and Azerbaijan, Israel and Palestine, Serbia and Kosovo, and Russia and Ukraine. To be sure, a few territorial contestations with less overt ethnocultural roots are included in some inventories of contemporary entrenched interstate territorial conflicts (e.g., Western Sahara, Iraq–Syria, Ethiopia–Sudan). Moreover, it is hard to identify any territorial disputes that are easy to solve. As Franck Billé persuasively argues, when maps are widely disseminated showing what a state sees as its rightful territory—a common practice in all cases—the effective dismemberment of a chunk of that territory carries with it damaging psychological impacts (what he terms *phantom pain*).[45] Nonetheless, the broader contours of the global territorial-conflict landscape clearly suggest that conflicts with nationalist ethnocultural foundations have the greatest propensity to inflict pain.

The Russian invasion of Ukraine in February 2022 offers recent confirmation of how significant an ethnoculturally rooted nationalism is when it comes to interstate territorial conflict.[46] On the eve of the invasion, Russian President Vladimir Putin delivered a major address to the Russian people in which he repeatedly emphasized the ethnocultural continuities between Russians and Ukrainians—arguing that they are essentially one people with a long, shared history and that Ukraine is effectively a region of the Russian nation-state.[47] Putin emphasized strategic considerations as well, of course, but his repeated references to ethnocultural history stand as a blatant reminder of the great psychological-emotional weight that an explicitly ethnocultural form of nationalism carries with it in the territorial arena.

Beyond the specifics of individual cases, there are important conceptual and practical reasons to pay more attention to the impacts of different types of nationalism and associated territorial imaginaries on interstate territorial conflict and its settlement potential. Conceptually, such a focus responds to the call for more consideration of the emotional-cum-psychological realm in the social sciences.[48] It also articulates with an emphasis in the emerging field of "critical border studies" on the political-economic *and* sociocultural circumstances that affect how borders are understood and function.[49] Moreover, it speaks to Radcliffe's call to consider how "different imaginative geographies . . . are drawn upon, re-articulated and negotiated, not only in periods of war, but on a day-to-day basis."[50]

In a more practical vein, the present study points to a potentially fruitful way of grappling with the disconnect between the cartographic representation of theoretically sovereign spaces and underlying political-territorial imaginaries. If, as argued here, nationalist arguments with an ethnocultural foundation are particularly ingrained and resistant to compromise, then insights into the geography of interstate territorial conflict can come from (1) juxtaposing the contemporary world-political map with maps showing the maximum historical-territorial extent of the states shown on the current map or their political-territorial antecedents,[51] and (2) in those cases where that juxtaposition reveals significant differences between contemporary and historical-territorial domains, distinguishing cases where one or more of the involved states has successfully cultivated territorial imaginaries grounded in a deep historical sense of ethnocultural distinctiveness from those that have not. To be sure, these approaches cannot be expected to produce definitive insights into which territorial conflicts are easier or more difficult to resolve; after all, just because a conflict is rooted in deep-seated ethnocultural territorial imaginaries does not mean it cannot be resolved (consider the effective resolution of long-standing territorial tensions between France and Germany and between Russia and China in the face of particular geopolitical and institutional circumstances). Nonetheless, focusing attention on the ideological foundations of national and territorial imaginaries can serve to direct attention to places where conflict resolution has a greater or lesser potential for resolution.

The Greece–Turkey and Ecuador–Peru examples discussed here make the point. To be sure, there are limitations to how much can be made of the contrasts between two cases, but the types of nation-building narratives and associated territorial imaginaries at play in those cases are found in many others. States such as Japan, China, Russia, and Iran developed in a cultural-geographic environment that allowed them to create and nurture a founding vision and associated territorial imaginary based on the idea that the modern polity is the homeland of a distinct cultural-historical nation. No conceivable counterpart could have developed in Indonesia, Angola, Gabon, or Afghanistan. Paying attention to such differences, along with the historical ideas and circumstances that gave rise to them, has the potential to deepen our understanding of the sovereignty challenges that are disguised in the typical contemporary world-political map, with its deceptive depiction of neat border lines separating states.

Jordan Branch

Reconceptualizing the State and Its Alternatives

Ideas, Infrastructures, Representations

In the enormous literature on the nation-state—its nature and characteristics, its origins and persistence, possible challenges and alternatives to it—the state as a concept has been defined at length. Even within a single academic field, such as political science, definitions vary widely: Is the state an actor, a formal institution, a network of relations, or something else? Rather than proposing a new, competing definition, this chapter instead aims to synthesize conceptual elements from across discussions, identifying how certain threads have constituted the state as today's dominant organizational form—including, in particular, its territoriality. Specifically, this chapter proposes reconceptualizing the state as a combination of particular ideas, infrastructures, and representations: *ideas* about organization, authority, and action; material *infrastructures* of control and communication; and *representations* in visual, linguistic, and other forms.

I am not arguing that this definition should replace existing concepts. Instead, this chapter suggests a thought experiment: *What if we reconceptualize the state in this more expansive way?* How might this suggest different answers to questions about the emergence, nature, and persistence of the state? What new questions would this generate, about the state and about the relationship among political ideas, material infrastructures, and—particularly relevant to this volume—visual and other representations? Most concepts define the state in terms of ideas alone, although often implicitly. Studies that have discussed infrastructural or representational elements have incorporated them as causes, effects, or tools of states or state building. Instead, this chapter proposes that we think of the state *itself* as a combination of ideas, infrastructures, and representations.

This reformulation draws on disciplines that have rarely been in conversation, and provides new means to address the emergence and contemporary trajectory of statehood. For example, widely varying explanations for the origins of "the state" tend to be framed as directly competing arguments.[1] Yet many are actually explaining different aspects of statehood: states' territorial and boundary-defined nature, their centralization, or their enormous capacity for intervention into society, and so on. Instead, by explicitly identifying diverse elements that together constitute statehood, we can clarify the distinct trajectories and drivers of each. This also allows us to rethink the "challenges" faced by states today: Which elements of the state do those challenges actually threaten, and which do they leave intact? This helps explain the persistence of the state in spite of the enormous transformations of the last half-century—including many that have been hypothesized to undermine the purpose, legitimacy, or practical functioning of states. Similarly with the potential for alternative forms of political organization or identity: we can rethink how likely or even possible various supranational, subnational, or nongeographic organizations are, given that a successful alternative to the state would need to be constituted by a combination of new ideas, infrastructures, and representations that are as powerful and (perceived to be) as effective as those that make up the state.

This chapter proceeds by, first, laying out some existing conceptual frameworks that suggest redefining statehood as a combination of elements. Next, it discusses each piece in turn—ideas, infrastructures, and representations—exploring how scholarship across disciplines has pointed toward these specific elements, even if they have not been drawn together in this way. Then, the chapter discusses how this particular conceptualization can provide new insights—or at least raise new questions—about the emergence of the modern system of territorial states, its persistence in the face of purported challenges and failures, and the possibility of alternative forms of organization.[2]

Redefining the State

Defining states as the combination of diverse ele-ments draws on existing arguments holding that social and political organizations or institutions can be constituted by disparate components. Prominent among these are the concepts of *assemblages* and *actor-networks*,[3] both of which emphasize particular ways in which diverse elements function as a whole. For example, modern states can be considered "the outcome of actually performed assemblages of all those practices of building it, protecting it, governing it and theorizing about it,"[4] or the effect of the "mechanisms and methods" of governance.[5] Patrick Carroll's study of how the modern state was "materially engineered"[6] defines the state as a combination of elements from three categories: ideas and discourses, practices and organizations, and territory, infrastructure, and population.[7] These come together to form a *plexus*, an "intertwined or entangled mass forming a complex web or network" with "multiple discrete forms connected together in complex ways."[8] Large institutions like the state, in other words, are the complex combination of diverse elements, joined in various networked relations.

Drawing on these frameworks, this chapter emphasizes the interplay of material and ideational factors, shifting away from the traditional privileging of ideas and rules as what the state "is" and material tools and representational practices as what a state "does." In other words, most studies of the state that do incorporate infrastructures and representations present those elements either as *tools* of state power; as *causes* of state formation, persistence, or failure; as *signs* or observable measures of the state's existence or features; or as the *effect* of the state and its actions.

Instead, this chapter argues that the state *is all three*: ideas, infrastructures, and representations.

Ideas

The first element is the *ideas* about organization, authority, and interaction that states embody, or are expected to embody. Much of the existing discussion in political science has focused here, on the history of sovereignty as a concept, how society and state are understood to be distinct, and the shift over time to a purely spatial-territorial idea of how political claims are separated.

The emphasis on ideas is apparent in the oft-cited definition of the state from Max Weber: the state is "that human community which (successfully) lays claim to the *monopoly of legitimate physical violence* within a certain territory."[9] Although physical violence is prominently mentioned, the key elements differentiating states from other organizations are ideational: the claimed legitimacy of the use of force, the monopoly of those claims (i.e., sovereignty as final authority), and the territorial limits of those claims. According to this definition, the combination of these features is what distinguishes the ideal-typical modern state from earlier forms of centralized political authority.

Distinguishing among the different histories of each ideational component of modern statehood has been a productive step in existing scholarship. For example, studies in the history of political thought have examined the emergence of particular concepts with new vocabularies and their deployment in texts and rhetoric.[10] On the international dimensions of modern statehood, one study considers how to identify the presence or absence of sovereignty, even in settings where the word itself did not exist.[11] And Stuart Elden has traced in detail the emergence of the concept of territory—in particular, the boundary-defined territory of modern states.[12] Each has different origins, with some emerging much earlier than others. For example, in the European context, the basic concept of sovereignty as recognizing no higher authority can be traced to the late Middle Ages, when legal theorists applied the Roman concept of empire to kingdoms: the king as emperor in his kingdom (*rex in regno suo est imperator regni sui*).[13] The spatial aspect of statehood is more recent, especially the exclusive claims over boundary-defined territories, which consolidated as an idea only in the seventeenth century.[14] Finally, the expectation that states have a capacity to intervene in society emerged in the nineteenth century, and has continued to evolve since then.

Although the question of state capacity may appear to be about the effects rather than the ideas of statehood, the centrality of ideas is demonstrated by the notion of a "failed state."[15] States described as "failed" in the post–Cold War era have fallen short of "full" statehood in different ways: some no longer exert authority successfully over the entirety of "their" territory, while others face no competitors within their boundaries but do not have the expected type of state capacity. In most cases, these "failed" states actually exhibit far more capacity and authority than many historical-political organizations, before the consolidation of the modern state. Today, in other words, both statehood and its failure are defined in part by these *ideas* about what a state is supposed to be.

Infrastructures

Yet ideas alone do not make the state. For one, those ideas are put into practice by infrastructures. These material systems do not merely provide tools for the institution of the state to act, nor are they only the material conditions of possibility of statehood. Instead, certain infrastructures can themselves be considered a component of statehood, just as much as the ideas defining sovereignty, legitimacy, or territory. This draws on a broad reading of arguments from science and technology studies (STS), such as Langdon Winner's position that some technologies "have" a politics.[16] Existing applications of STS to the study of political institutions have highlighted the ways technologies are *similar* to institutions politically: both are created by agents but later shape the context for action.[17] Here, I am arguing instead that institutions *are* in part infrastructural, or technological.

Consider again the concept of state failure. As noted, this is usually framed in terms of a failure to live up to the expected features of statehood, defined in ideas. For example, studies of areas of "limited statehood" examine deficits in legitimacy, defined as "the population's sense of obligation or willingness to obey the authority."[18] This certainly captures part of the nature of state failure or weakness, but it misses another, essential element. As Peer Schouten

argues provocatively, state failure may be as much about *material* shortfalls as it is about the failure of a social contract between ruler and ruled.[19] Thus state failure involves both a weakened social contract *and* the absence, removal, or destruction of material infrastructures.

This builds on studies that have posited various roles for infrastructure in statehood. For one, Michael Mann's history of political organization proposes a distinction between the "despotic power" of a ruler to make commands across a range of actions and the "infrastructural power" comprising "the capacity of the state to actually penetrate civil society, and to implement logistically political decisions throughout the realm."[20] While this stays largely in the realm of ideas defining the state, it incorporates state capacity and discusses the importance of infrastructures as the "routinized media through which information and commands are transmitted."[21] Similarly, the concept of "logistical power" posited by Chandra Mukerji emphasizes "the ability to mobilize the natural world for political effect," particularly through large-scale infrastructural projects.[22] Jo Guldi's study of the British "infrastructure state" explores how "political activity is engendered by changes in infrastructure" like road networks.[23] None of these, however, makes the essential point of this chapter, that infrastructures may be as much a component of statehood as ideas are.

Actor-network theory (ANT) and related approaches offer more direct support for the argument that the state itself is composed of infrastructures. ANT emphasizes the interweaving of material and nonmaterial, of human and nonhuman, in the networks that constitute both actors and the outcomes of interaction.[24] As Bruno Latour argues specifically about the emergence of modern states: "The 'rationalization' granted to bureaucracy since Hegel and Weber has been attributed by mistake to the 'mind' of (Prussian) bureaucrats. It is all in the files themselves."[25] This chapter makes a more limited—and synthetic—argument, holding that this "rationalization" and other characteristics of the modern state are *both* in the minds of actors *and* in infrastructural systems like material files. Carroll comes closest to this approach, proposing that the modern political world is composed of "engineering states": "sociotechnical systems, ingenious contrivances of things human and nonhuman."[26] The nonhuman side of this includes both "objects of engineering culture—land, roads, buildings, and so on"—as well as tools of measurement, representation, and communication.[27]

Modern states are particularly infrastructural, as is modernity writ large.[28] Yet *all* polities have had an infrastructural component. Political organization has always been defined, at least in part, by infrastructures of control, including information systems (defined broadly). Taxation requires recordkeeping, which relies on some form of material information storage and retrieval. James C. Scott's study of "early states," for example, highlights the role of measurement and writing, instantiated in material systems: "The creation and imposition of a written code throughout the city-state replaced vernacular judgments and was itself a distance-demolishing technology."[29] The subsequent increases in hierar-

chical complexity and scale of polities in the ancient world were paralleled by developments in various forms of information storage and communication, from new written media to faster and denser courier networks.[30]

The specific infrastructures of the modern state include information, communication, transportation, and bordering technologies. In terms of information systems, early modern states relied on increasingly complex recordkeeping, which expanded enormously during and after the nineteenth century. That period also saw the development of the electric telegraph as the first means of separating the communication of information from its material storage (and thus from transportation). States today are constituted by an enormous network of communication and informational systems, for recordkeeping, for intelligence and surveillance, and for governing from bureaucratic centers over peripheries.

Polities have also relied on transportation to exert control, particularly on scales larger than a single city or local region. Returning to the Weberian definition of the state, the *successful* claim of a monopoly on the legitimate use of force requires the ability to do so at a distance. Road networks, ship technology, and later railroads and further mechanized transportation systems have been part of this infrastructure.[31] For states to exist and act, they require those transportation systems not as a tool of control, but as a component of control itself—in other words, as a component of the state. Luca Scholz (this volume, chapter 10) illustrates this well, revealing how the governance of roads needs to be included in how we represent the

Holy Roman Empire. Those roads and their control *were* part of the polities of the empire, not just a tool of those polities or an effect of their rule. Similarly with Franck Billé's discussion (this volume, chapter 7) of the hub-and-spoke road networks in contemporary Russia: the radial, gap-filled sovereignty of the Russian state is not just represented by this road network, nor is that form of sovereignty simply an effect or cause of the structure of transportation. The state and its nature *are*, in part, the road network and its specific features.

Finally, modern state infrastructures are especially focused on implementing the ideal of linear borders. Although territorial ideas developed earlier, it was only when those borders were materially implemented that statehood took on its contemporary spatial form. The infrastructures of borders and bordering are varied: from walls and physical barriers that both mark and enforce a boundary[32] to systems of identity and control such as the modern passport regime.[33] Incorporating the infrastructural side of bordering helps identify the complexity of these practices, as many states have shifted their border controls away from the linear border itself, both internally and externally. In the US case, this ranges from the installation of US border controls at airports outside the United States to internal identity checks far from the physical border.[34]

Existing arguments often focus on the *effects* of these technological systems on state building. This chapter, again, suggests shifting the focus to how these systems themselves are part of the state. Especially when we think about the category of border infra-

structure, it is hard to conceptualize the territorial nature of the state without including these material systems. The *idea* of clearly delineated territories alone is not what shapes international politics; what matters is the combination of the hegemonic status of that idea (states are supposed to be defined by boundaries) with the ways the idea is instantiated materially—and with representations, discussed next.

Representations

The third element of statehood consists of the *representations* that embody the state and that allow actors to make sense of what the state is or should be. The state as an idea is unobservable, and the material infrastructures of the state are vast and difficult to grasp in their entirety. Thus, when actors think about, discuss, and enact the state, they deploy a host of representations—from visual imagery in mapping, to linguistic terminology, to written legal descriptions.[35] Given that these representations are central to statehood, this chapter, again, proposes reconceptualizing them as part of the state, rather than a cause of state formation or an effect of state action.

Existing research on the emergence of the state has, in fact, incorporated representations. In earlier work, for example, I argued that the specific territoriality of the modern state, with its focus on discrete boundaries and homogeneous spatial claims, was shaped and made possible by developments in mapping technology.[36] Here I push this connection further, moving beyond the argument that "maps *made* the state," so to speak, to argue that mapping *is* (part of) the state. Thongchai Winichakul's work on national identity in Siam/Thailand can be read through this lens: the country's "geobody" did not exist so much in demarcated boundaries on the ground, but instead in a variety of map representations, from careful surveys to "logo maps" of the country's outline. Or consider Timothy Mitchell's study of late nineteenth- and twentieth-century Egypt.[37] As land was mapped in increasing detail for taxation, the site of governance moved from the land itself to the "map room," where the representational artifacts (i.e., maps) were stored, examined, and measured.

Several chapters in this volume can be read in similar ways. Barbara Mundy (chapter 8) suggests the importance of representations when highlighting the transfer of Aztec maps to the Spanish as part of a shift of sovereign authority. Valerie Kivelson (chapter 1) notes the importance of representations for the Russian Empire circa 1700: maps produced in Russia incorporate and instantiate the empire's lattice-like sovereignty. And Ali Yaycıoğlu's discussion (chapter 4) of the Ottoman refusal to use one form of representation (European maps) in favor of another (verbal discussions and written documents) illustrates how the Ottomans instrumentally rejected the form of representation used by their adversaries. These examples reveal how the territorial aspect of modern statehood involves all three elements: ideas about how to define political rule spatially, infrastructures of boundary control and creation, *and* visual representations of territory. Without those representations, the institution of the modern state would not have some of its essential characteristics.

While mapping is central to the territorial nature of statehood, other representations also constitute the state. For one, how the state and its core features are measured, catalogued, and described are fundamental. Eighteenth- and nineteenth-century statistics and censuses, for example, came to define a state's population similarly to how maps defined its territory.[38] Again, these representations are a component of the state rather than a tool of state power: as Bruce Curtis puts it, "Census making is itself a practice of state formation."[39] In other words, the state is not only strengthened by the census, the state *is* the census. As with mapping, this representational element also involved the creation of an enormous material and bureaucratic infrastructure.

Second, at the level of international law, representations in legal conventions and concepts constitute a part of what makes a modern state. For example, Lhamsuren Munkh-Erdene (this volume, chapter 2) discusses the Great Code, adopted by rulers of Inner Asia in 1640 to create a "confederation of principalities." This code served as a core representation of political organization in the region, defining and instantiating the idea of independent but collectively secured principalities. In addition, Yaycıoğlu discusses the principle of *uti possidetis*, a diplomatic principle of territorial possession at the conclusion of conflict. This is both an idea and something that becomes instantiated in representations, in the case of Ottoman–European negotiations, in written treaties. Today, the 1933 Montevideo Convention posits that "the State as a person of international law should possess the following qualifications: (a) a permanent population; (b) a defined territory; (c) government; and (d) capacity to enter into relations with the other States."[40] For the international dimensions of sovereignty and statehood, this legal representation is essential—what is and is not recognized as a state is part of what constitutes statehood.

Rethinking the State's Emergence, Persistence, and Alternatives

How does this new concept of statehood help? For one, it pushes against an unproductive reading in political science of how technologies and politics interact. These arguments, often implicit, see technologies dichotomously, either as simple tools or as deterministic constraints on action.[41] In the context of statehood, that dichotomous interpretation has boiled down to asking whether the state is "caused" by new technologies or whether it creates and uses those systems for its own ends. If, instead, we conceptualize those systems as part of the state itself, this would reframe questions in three areas: the *emergence* or origins of statehood; the *persistence* of states in spite of numerous contemporary challenges; and the nature of historical, existing, or hypothetical *alternatives* to the state.

Looking for single origin of "the state" is impossible, even if we were to maintain the conventional focus on ideas.[42] Including infrastructures and representations further highlights the diverse histories of the state's components, allowing us to trace when and how each emerged and came to be embedded in the state. Then we can parse out the relationships

among the components of statehood within the evolution of this institution over time, seeing each take on its modern form by different logics. In other words—and, admittedly, contrary to how my earlier work framed this question[43]—we could move away from asking if certain representations or infrastructures were drivers of state formation and instead ask how they have constituted one part of state formation itself.

For example, a number of works have traced some aspects of the European state to the late Middle Ages.[44] Yet the specific features of states examined are almost entirely in the realm of ideas: ideas about the sovereignty of monarchs versus the papacy, for example. While modern states are in part built on those ideas, statehood is equally reliant on developments in infrastructural capabilities and representations. This supports recent work reconsidering the history of "absolutism," for example. By bringing in infrastructures and representations *alongside* ideas, we can see how absolutism, like other state forms, has *both* an ideational history and a practical history. While some of the core ideas underlying absolutism certainly appear in the sixteenth and seventeenth centuries,[45] scholarship has "reveal[ed] the real limitations of absolute monarchy even in those countries where it was not subject to major challenges."[46] The existence of state infrastructures strong enough to impose absolutist ideas throughout a territory appeared only in the nineteenth century.[47] And the interplay between absolutist ideas and representations like mapping is also complex, and prolonged.[48]

This concept can also shed new light on the challenges to statehood posed by contemporary globalization—or at least bring a fresh perspective to the discussion. Much of this debate has started by identifying particular features of contemporary politics, society, or economics that undermine or threaten the state, framed largely in terms of *ideas* of statehood, such as the claim to exclusive control within boundaries. Thus "boundary-crossing" or "boundary-erasing" threats undermine a state's claim to control its territory; or changes in the logic of collective action due to the increasing scale of market activity undermine the effectiveness of the state as a political means to manage those activities.[49]

Yet the state, mysteriously, persists. This becomes less mysterious when we think about the state as composed of more than ideas. For example, while challenges may undermine the idea of exclusive, centralized authority—in terms of legitimacy or national identity, for example—a more complex picture emerges when we include infrastructures and representations. Certain infrastructures of statehood are threatened by globalization, but others are strengthened by those same forces. For example, while the internet has eroded the ability of some states to control information within their borders, it has also given strong states new opportunities to exercise informational surveillance and control.[50] Moreover, Carroll highlights that the supposedly postmodern nature of contemporary politics and statehood continues to rest on a persistently "modern" material basis.[51]

An even greater disconnect exists between representational aspects of statehood and the weakening ideas of the state. Supposedly state-undermining globalization processes have not created or supported new representations that are able to challenge the territorial state in maps, language, or international law.[52] There are few if any representations of alternatives to the territorial state that are as evocative and convincing as the mapped territory. Most of the visual representations of challenges to statehood continue to depict the world as composed of boundary-defined states. Consider, for example, maps of cybersecurity attacks, often understood to be one of the novel threats *least* amenable to state control via territorial enforcement at boundaries. These maps often layer the novel element on top of a standard political map of state boundaries;[53] likewise with maps of low-tech but purportedly state-threatening phenomena today, such as mass migration. Maps of recent migration "crises" have almost invariably built on—and reinforced—the "grid" of state boundaries.[54] Thus the state survives, as its representational element remains strong, perhaps even hegemonic, as a means of depicting and thus understanding the political world. The state persists not *because of* the hegemony of the political map of the world; instead, the state persists *in* that representation itself.

This shift in emphasis away from focusing exclusively on the ideas that are undermined or threatened also helps us reframe how we think about "exceptions" to statehood: failed states, special economic zones, nonstate groups that control territory, and so on. Because of the persistence of state-affirming infrastructures, as well as the continuing use of boundary-filled representations, those exceptions are discussed and treated as exactly that, as *exceptions*. Few political actors want something other than their "own" state, as most are aimed at either gaining control of an existing state or carving out—territorially—a new state. The Islamic State (or Daesh) was a rare exception to that trend: a group that controlled territory, made use of and even constructed governing infrastructures, but seemed to have no interest in achieving international recognition or satisfying the conventional representational requirements of states: UN membership, negotiated and mutually agreed-on boundary delimitation, and so on. It is suggestive that so few other "challenges" to states take this approach, which would represent a more significant threat to the entire state system than does the more common separatist goal of becoming a state.

Finally, this concept helps explain the absence of significant alternatives to territorial statehood— that is, organizational forms that could directly replace the state. For one, it is difficult to imagine an alternative that would operate as successfully across all three elements, particularly in terms of representations as evocative and naturalized as the mapped state.[55] How could we represent new possibilities for political organization in a more convincing fashion, in a world "seduced" by the nation-state map? Part of why the existing way of mapping became so closely integrated with the state was because it was taken up for a variety of purposes, motivated by commercial, scientific, and artistic as well as political goals.[56] New representations thus need to be more than a novel way to depict a nonstate form of organization; they also need to be convincing and appealing enough to be incorporated into how an organization is identi-

fied, and how it identifies itself. This is not impossible, of course: other contributions in this volume suggest some possible pathways to those novel and evocative depictions.

There is the related possibility that different forms of organization or mobilization might emerge that, even if not competing directly with the state, might be layered on top of an enduring state system—a possibility this chapter's concept helps situate as well. Digital information networks, for example, have their own definitional ideas, infrastructures, and representations. In the case of the internet, the ideas and infrastructures can be readily identified: ideas in the form of ideologies, protocols, and code and infrastructures in the private and public hardware for information storage, retrieval, and transmission. Yet representations of the internet vary widely, and efforts to map the internet visually are far from settled on a single hegemonic form (like the nation-state map has done).[57] Most of the politically consequential representations of the internet are in language rather than visual depiction, in efforts to describe the nature of global networks with analogies and metaphors. For example, US government and military agencies use the term *cyberspace*, a term with embedded metaphors that have rhetorically supported expanding the role of the military in managing the internet.[58] Searching for possible alternatives to the state may require us to shift away from the focus on visual representations that the mapped nation-state relies on.

The absence of readily identifiable alternatives to the state, viable across all three dimensions, does not mean that the state is immutable. Tensions and contradictions between the three elements may eventually yield significant change in the institution. Thus, while today the dominance of the mapped image of the state supports the persistence of the institution, the tension over time between that representation and the increasingly complex infrastructures of governance—as well as weakening ideas of territorial sovereignty and national identity—could be a productive place to look for future change. In other words, the three elements of the modern state do not inevitably have to appear together, nor will they reinforce one another indefinitely.

Conclusion

The goal of this chapter has been to propose a new way to conceptualize the state, and to explore some ways this new concept could be useful for understanding the historical and contemporary trajectory of statehood—particularly the persistence of the state in spite of its many "failings." Including infrastructures and representations incorporates some of the modern state's strongest and most enduring elements, ways it has *not* failed: material infrastructures of control and communication, representations that reify the boundary-defined territorial state as a natural political form, and so on.

One extension of this concept could be in applying it to institutions beyond the state. How do other large-scale institutions function in terms of ideas, infrastructures, and representations? The United Nations, for example, is both a collection of ideas (goals of cooperation and peaceful conflict resolution as well as rules about membership, voting, and decision-making) and a host of infrastructures and representations. The UN flag, for example, was explicitly designed to embody and reinforce those ideas.[59] Informal institutions or international regimes can also be understood through this lens. Global trade, for example, rests not only on a collection of ideas about liberalization, reciprocity, and comparative advantage but also on the material infrastructure of containerized shipping and the linguistic representations of trade as an inevitable feature of modernity. Even when some of the core ideas of liberal trade have been threatened or undermined—as with shifts toward explicitly protectionist rhetoric and tariff policy by the United States—the material infrastructures of trade persisted until faced with COVID-19-era supply-chain disruptions.

Examining historical-political organizations could also be productive, and could helpfully move us away from the all-too-common framework of looking for something "state-like" in earlier eras.[60] Instead, we can ask what ideas, infrastructures, and representations of political rule existed, identifying some elements that likely resemble statehood (centralization of decision-making in many contexts, for example) and others that do not (such as the almost complete absence of visual representations of many historical-political forms). Reframing historical models in their own terms, rather than as something falling short of statehood or having a lineage to statehood, may allow for better understanding of our contemporary situation as well.

Franck Billé

Voluminous, Scattered, Distorted

On the Limits of Cartographic Representations

We all share the sense that we live in a transformative moment of territoriality. The ground is shifting under our feet.

—CHARLES S. MAIER, *Once within Borders*

At the end of the Cold War, there were fewer than a dozen border walls worldwide. There are seventy-four as of this writing, with at least fifteen others at some stage of planning. This by-now somewhat hackneyed opening sentence—variations of which are used in countless books on borders—remains very potent as a symptom of global forms of nativist populism and growing economic disparities. Beginning after the fall of the Soviet Union in 1991, the proliferation of border walls has increased further since, with the rise and consolidation of autocratic regimes in Europe, China, Russia, Turkey, India, and elsewhere.

The COVID-19 pandemic has brought further national retrenchments, turning most states into wannabe-autarchic entities, including within the European Union. The fact that barriers were erected primarily along international borders speaks to a potent spatial imaginary where national categories continue to mold our sense of safety and danger. As northern Italy was suddenly transforming into the principal site of COVID-19 infection in Europe—prompting several countries, including the United States, to issue advisory restrictions for the whole of Italy—countries bordering northern Italy were largely absent from news headlines. In an echo of familiar weather-map visuals where atmospheric conditions appear to be neatly bounded by state outlines, Italy was presented suspended in the air—virally tied to distant Wuhan and Iran, while entirely detached from its geographic context.

And yet, if the idea of the state as a discrete, bound unit remains very potent, the planet's "mosaic of distinct states has melted into . . . a scattershot of statelets, militarized cities, and transnational flows."[1] Our familiar jigsaw-like political geography seems increasingly detached from the realities of border management, with a growing abundance of "empirical spaces of legal suspension."[2] Recent literature on migration has drawn attention to the cross-state schemes seeking to keep unwanted migrants at bay. The United States, the European Union, Canada, and Australia in particular have developed strategies to block, funnel, and filter outsiders—outsourcing operations, setting up offshore naval bases or territories, and even weaponizing terrain itself. The border line itself has turned blurry, while border crossings of all kinds have become a staggered process involving airline check-ins, points of embarkation and disembarkation, and transit areas before the actual border is reached.

To account for the nonlinearity, nonhomogeneity, and noncontiguity of political borders, terms such as *unbundled, flexible, aleatory, paternal, graduated,* and *variegated* have been coined by social science and political scholars as hyphenated qualifiers of sovereignty. While these variants of modified sovereignty are useful to unpack the spatial formation of neoliberal economies, interpreting territorial discontinuities as exceptions reinforces in effect the Westpha-

lian model as the norm.[3] These spatial and temporal exceptions in fact function as instruments designed to work within the agreed system without having to challenge its core tenets. Instead of actual exceptions, then, they constitute internal devices inherent to the inner workings of territorial sovereignty—not a territorial ambivalence that exists *in spite of* the state, but one that is *structurally folded within it.*

The very fact that we continually refer to them as exceptions is at the core of the argument in this chapter. It speaks on the one hand to our attunement, as scholars, to the entanglements and spatial complexities of a modern state that is always less and always more than what appears on the map. But it also betrays the enduring visual force of political cartography, of a neat arrangement of discrete entities with no gaps or overlaps—ultimately indexing a lack in our ability to imagine other forms of spatial belonging. These two somewhat contradictory views—the Westphalian model of territorial sovereignty is one we are affectively invested in, yet one we recognize as a myth—constitute what I describe here as a breach. As our economic, cultural, and political world grows more and more entangled, ideals of discreteness get increasingly accentuated in the form of border walls and populism. We live, writes N. Katherine Hayles, at a time "when the planetary cognitive ecology is undergoing rapid transformation, urgently requiring us to rethink cognition and reenvision its consequences on a global scale."[4] Yet we do not seem to be able to think beyond the existing model, flawed as it is. Even though trade and migration perforate borders, Benjamin Bratton writes, state sovereignty and supervision over information flows are nonetheless dramatically reinscribed and reinforced.[5] He adds: "Instead of lamenting all the exceptions to the norm, hoping that they will get back in the box where they belong, perhaps it is time to map a new normal."[6]

Critical cartographers and historians have convincingly argued that the map preceded the territory, and they have shown that imperial maps in particular played a key role in colonizing new territories by portraying, ahead of actual exploration, entire continents as blank spaces or, alternatively, as places inhabited by monsters or fantastical creatures.[7] We are now witnessing a critical reversal. The territory has largely caught up with the map, and it is the map, having lost its earlier promissory role, that is now struggling to keep up with the territory. In this paper, I examine the limits of cartographic representability regarding the three core tenets of territorial sovereignty—containment, homogeneity, and contiguity. Exploring each of these in turn, I suggest in the conclusion that our attempts at representation of national territory may need to be jettisoned altogether in order to open up spaces of belonging organized in alternative ways.

Tenet 1: Containment

Donald Trump repeatedly claimed on the campaign trail before his election in 2016 that "we don't have a country if we don't have borders"—a statement that clearly resonated with voters.[8] Regardless of political orientation, the significance of national borders is deeply ingrained in modern political practices

to the point of having become a self-evident truth. The fact that state sovereignty is conceptualized as homogenously distributed has brought about an emphasis on edges, thereby turning on its head the previous conceptualization of the national space as radiating outward from a center. Disputes over minuscule, economically insignificant, and frequently uninhabitable pieces of real estate are thus a modern phenomenon that would have been incomprehensible before the advent of the map.

A good example of such a dispute is the Sino–Indian border across the Himalayan range over which India and Pakistan have lost more than two thousand soldiers since 1984—the vast majority killed by the weather and the terrain. This region was left unmapped at the Partition of India, as neither side anticipated it would become a matter of contention. The situation is evolving, however, and there is a palpable sense that the Indian and Chinese sides are creeping toward full appropriation of the border with the aid of new technologies such as laser fences, motion sensors, CCTV cameras, and a network of radars, and that the gap will eventually disappear.[9] This represents a classic case where the map has anticipated the territory and where technology is attempting to reify these representations. It echoes border consolidations elsewhere in the world and suggests that this process of containment is far from complete.

Perhaps more telling are the ways in which the map is also being modified in line with changes in the geography. Italy's northern border with Austria and Switzerland follows the watershed that separates the drainage basins of Northern and Southern Europe.

Running at high altitudes, the border crosses snow-fields and perennial glaciers—all of which are now melting as a result of anthropogenic climate change. In 2005, with borders no longer stable on the ground, "Italy and Austria signed into law a bilateral agreement, which introduced the definition of a 'moving border' for those cases where the line is subject to 'gradual natural processes'—that is to say, on glaciers."[10] A grid of twenty-five solar-powered sensors has been fitted on the surface of the glacier at the foot of Mount Similaun, and every two hours these sensors record data, allowing for an automated mapping of the shifts in the border.

In these two very different border contexts, technology aims to close the gap between physical reality and representation, between map and territory. The promissory role of sensors tracking borders in real time, or of drones surveilling spaces beyond human reach, assuages cartographic anxieties and maintains the totalizing fiction of the nation-state in suspended belief. The more precise and sophisticated these technologies, the more ontologically secure the borders. Ever more detailed and accurate, inching toward an isomorphic relation between the map and reality, these attempts, however, betray an imaginary of the territory that is two-dimensional.

As an example, a few years ago, a British company divided the world into a grid of 3-by-3-meter squares and assigned each one a unique three-word address. Having identified a persistent lack in existing locating systems, the company, what3words, can pinpoint an exact position and direct a user to it—a useful advantage for large buildings or campuses with multiple entrances, or to locate a lost hiker in a remote park. Undeniably valuable, the system was in fact adopted officially by the postal service in Mongolia, a country with little infrastructure and no street addresses—though this mapping system, tied to the horizontal plane, is woefully inadequate to identify elevation.

In recent years, though, forays into volumetric spaces have become increasingly routine. The definition and contours of volumetric sovereignty are in constant evolution, partly through the possibilities afforded by new technologies, but partly because these are relatively new questions. The notion of airspace sovereignty was recognized just after World War I as a response to aerial reconnaissance and bombings, while the concept of "territorial waters," in existence since the late eighteenth century and generally limited to three nautical miles (the typical range of cannon fire from shore), began to be challenged in the 1940s. More recently, the notion of a continental shelf (defined as the shallow seabed that extends from the part of the shore permanently submerged down to the continental slope, typically at a depth of one hundred to two hundred meters) has provided coastal states with further possibilities to extend their sovereign rights beyond their exclusive economic zone (EEZ) in ways that are complex, nonlinear, and incomplete.

Especially significant is the materiality of that space beyond the terrestrial—air, water, ice—which challenges traditional forms of containment. Yet incursions into those realms are often conceptualized and managed through a land-based imaginary. In 2016, the Ministry of Foreign Affairs of Japan

FIGURE 7.1. "Do You Know the Shape of Japan?," 2016. Courtesy of the Ministry of Foreign Affairs of Japan.

launched a visual campaign about its disputed borders. A poster titled "Do You Know the Shape of Japan?" highlights three points of tension: the Northern Territories, Takeshima, and the Senkaku Islands, disputed with Russia, Korea, and China respectively (figure 7.1). What is truly radical about this map is less the territorial claims that it stakes than the confident demarcation of a hydro-territorial entity, neatly defined against an undifferentiated background. The same outline had previously been published on the ministry's website; however, the online representation merely traced the extent of Japan's territorial waters. The poster's terraqueous map, by contrast, distin-

guishes between Japanese and non-Japanese waters and gives equal weight visually to both land and sea.

In his conclusion to *The Cartographic State*, Jordan Branch wonders whether new representations can be devised to analyze emergent pluriterritorial, polycentric, and multiscalar geographies of globalization, and the spatial remit of this question may be expanded here to include three-dimensional territorial incursions beyond the terrestrial. In fact, through explicitly pedagogical means, the Japanese poster seeks to do precisely that and to foster affective attachment to these more-than-human geographies. Branch suggests, convincingly, that alternative representations may prove too abstract to build "new notions of identity, community, or authority"—unlike classic political maps, "with their clean linear boundaries and homogeneously colored spaces."[11]

Tenet 2: Homogeneity

A second core tenet of modern territorial sovereignty is that it be homogeneous. A much-cited statement by Benedict Anderson sums it well: sovereignty should be "fully, flatly, and evenly operative over each square centimeter of a legally demarcated territory."[12] Cartographic depictions also suggest territorial sovereignty be isotropic; in other words, that its physical properties have the same value when measured in different directions. This concept is closely related to the notion of homogeneity, but isotropy is more insistent on directionality—an important point to make, given that sovereignty was previously radiating outward from a center. Indeed, if modern sovereignty is homogeneously distributed, much stronger links still persist between the core and the periphery than between different points in the periphery.

This is particularly true in highly centralized states like Russia, where vertical links (to the capital and the president) are strong and horizontal ones weak. The vertical power relationships are reflected in the radial roads that link the metropolis to the regional capitals, then to the provincial centers, the district centers, and finally to the villages on the periphery. The only roads kept in good working order are those leading "upward." Meanwhile, the roads running "horizontally" between small towns or villages are neglected—often becoming impassable in bad weather—or simply do not exist. Similarly, bus services, cellphone signals, and so on fan out from Moscow and from each provincial center, but they become more exiguous and may give out at the edge of each jurisdiction. This creates pockets of remoteness in the gaps between the skeletal radial lines, which can sometimes be found even quite close to a major city.[13]

This situation contributes to what one might call Russia's "lumpy sovereignty," a national territory that also includes numerous gaps and denser nodes. There were far more of these during the Soviet period, when Russia was home to dozens, possibly hundreds, of so-called closed spaces, ranging from border exclusion zones to villages, sometimes entire cities, with links to the military-industrial complex, such as the port city of Vladivostok. In the many exclusionary spaces that still remain in contemporary Russia—notably, villages and settlements on the Russia–China border—restrictions continue to apply to both foreigners and nonresident Russians. For people residing in these exclusion zones, it can feel like being utterly cut off from the world: visits from outsiders are restricted and regulated, thus making it virtually impossible to, say, open a business.[14]

As is well known, Soviet maps were deliberately falsified for most of the history of the Union, misplacing rivers and streets, distorting boundaries, and omitting geographical features, on orders of the secret police. As the Soviet Union's chief cartographer acknowledged in 1988, "Roads and rivers were moved. City districts were tilted. Streets and houses

NEARLY **2 OUT OF 3** PEOPLE LIVE WITHIN THE 100-MILE BORDER ZONE

FIGURE 7.2. "The 100-mile border zone." Originally posted by the ACLU, https://www.aclu.org/know-your-rights/border-zone. Copyright 2024 American Civil Liberties Union.

were incorrectly indicated. For example, on the tourist map of Moscow, only the contours of the capital are accurate."[15] While these maps were inaccurate from a geographical standpoint, one might argue that post-Soviet maps, which indicate buildings and features in their actual location, are also inaccurate insofar as they fail to reflect the realities of territorial management—two neighboring settlements may appear similarly on the map, yet one might be closed to outsiders or unconnected by roads.

I am reminded here of Lauren Benton's point that European empires appeared on maps as territories in the same color as their metropolitan centers, yet imperial space was "a fabric that was full of holes, stitched together out of pieces, a tangle of strings . . . politically fragmented; legally differentiated; and encased in irregular, porous, and sometimes undefined

borders."[16] Benton suggests that a more accurate representation of imperial power "would show tangled and interrupted European-claimed spaces and would represent, perhaps in colors of varying intensity, the changing and locally differentiated qualities of rule within geographic zones."[17]

Her point remains largely valid, perhaps even more so in today's neoliberal landscape, including in federal (i.e., less centralized) and democratic (or at least democracy-adjacent) countries like the United States. The US border zone, for instance, is a thick ribbon, very much unlike the thin abstract line that appears on maps. It extends one hundred miles inland, thus completely enclosing a number of states such as Florida and New York, as well as substantial portions of the most densely populated states like California (figure 7.2). In fact, nine of the ten largest

US metropolitan areas fall within this area. Within the border zone, federal regulations give US Customs and Border Protection extended authority to conduct random and arbitrary stops and searches over, in effect, roughly two-thirds of the United States' population.

The situation in the United States is in fact very complex, with the existence of various kinds of territories, classified as organized or unorganized, incorporated or unincorporated—denoting varying degrees of integration, political representation, and citizenship rights. Consider people born in American Samoa, an unorganized and unincorporated territory: they are US nationals but not US citizens (nor citizens of *any* country).[18] From special economic zones and corridors to territories that states control but do not have sovereignty over, and from territories and dependencies where citizens have fewer rights to islands where the strict religious rules of the mainland do not apply, these are some of the countless folds and kinks in a state's spatial fabric imagined to be, and idealized as, homogeneous.

Tenet 3: Contiguity

Closely linked to containment is the importance of contiguity—specifically, that national territory should be, at least ideally, singular and unbroken. Like "containment" discussed above, this is a modern concept that would have held very little meaning prior to the treaties of Westphalia. Indeed, medieval Europe had a very different geography. Land titles were exchanged freely between noblemen,

resulting in numerous zones of exception, gaps, and enclaves. But by the eighteenth century, the geographer Alastair Bonnett writes, enclaves were being seen as a problem and the "rational world of the Enlightenment tried to sponge away the dark and unmanageable world of enclaves."[19] Of the enclaves still in existence in Europe, the most extensive and convoluted is the enclave complex of Baarle-Hertog and Baarle-Nassau, two towns enmeshed into a single urban entity across the border between Belgium and the Netherlands. On a total area of 32.35 square miles, the two towns are intertwined into each other. Situated only three miles away from the Belgian mainland, the Belgian town of Baarle-Hertog is made up of twenty-two exclaves, with the remaining territory around it forming the Dutch town of Baarle-Nassau. The situation is complicated further by the presence of seven counter-enclaves: plots of Dutch territory located within the Belgian enclaves. This territorial jigsaw is compounded by the irregularity of the shapes of the enclaves themselves, with border lines slicing haphazardly across fields, streets, office buildings, and private homes.[20]

If Baarle residents have quite successfully used these enclaves to market the town for tourist purposes, daily life in enclaves elsewhere in the world is often less benign. The largest and most intricate enclave complex, known as the *chitmahals*, was located until 2015 across the India–Bangladesh border. It consisted of 102 Indian enclaves on Bangladesh territory and of 71 Bangladeshi enclaves within India's mainland. This spatial enmeshment was complicated further by the existence of counter-enclaves and even—a unique case worldwide—a counter-counter-enclave: a tiny piece of Indian territory in a fragment of Bangladesh, itself within a piece of India within Bangladesh. Residents of these enclaves were often described as "stateless" since they lived in zones outside of official administration and officials of one country could not cross a sovereign frontier to administer territory.[21] After decades of attempts at a resolution, the governments of India and Bangladesh finally announced their intention to resolve the issue through a land swap, giving residents a choice of nationality. While the resolution was largely prompted by the considerable human suffering they caused as well as the difficulties for both states to administer such labyrinthine geography, there was also the sense that the enclaves were anachronistic and that both India and Bangladesh needed to "regularize" their border to become fully aligned on modern models.

In comparison, Baarle is more like a curiosity, within a European Union that is increasingly uniformly managed. Sharing a currency, a language, and most of their laws, the two towns offer their residents very similar living experiences. But even benign cases such as these can feel irksome. A few years ago, after a public presentation I had just given on the enclave complex of Baarle-Hertog/Baarle-Nassau, a member of the audience explained over drinks that the talk had been interesting but had made him angry. "Why do they still have this mess? Why can't they sort it out? I hate it." Clearly, enclaves jar with the cartography to which we have become accustomed, and there is a strong sense that these atomized geographies have no place in modern political geography.

Conversely, France and Texas appear to be two examples of prototypical logomaps. Both are used extensively as logos and as units of measure (such and such country being x times the size of Texas or France). They are both balanced in terms of weight, fitting neatly within a circle; much has been written in particular about France and its regular shape approximating a hexagon.[22] In *Cartophilia*, Catherine Tatiana Dunlop notes that the practice of superimposing a hexagonal outline over the relief map of France became commonplace in French classroom geography texts throughout the Third Republic: "The hexagonal form created a public perception of France's authentic national shape as preordained and almost God given in its mathematical perfection, giving the French a strategy for arguing against the ethnic and linguistic justifications for national borders that other Europeans had embraced."[23]

What these ideal representations leave out, of course, is France's vast former colonies and overseas "possessions." France's spatially neat hexagon is only the tip of the spatial iceberg that the country constitutes. Its numerous overseas territories make the country a truly global entity, with a territorial footprint accounting for 18% of the mainland's territory and an exclusive economic zone representing 96.7% of the total. Even fragments closer to the mainland and fully integrated in terms of sovereignty bear a fragile visual relationship to the state to which they're attached. Corsica, a few miles off France's southern coastline and usually included in the logomap, is absent from the graphic imaginary of the hexagon. Similar to diacritics, the relationship of fragments to their continental state feels additive and impermanent. A quick Google search for images of the "USA logomap" brings up outlines of the mainland without Alaska or Hawaii. In the rare instances where the two states are included, it often feels like an afterthought: reduced in size, they are tucked in where space allows.[24]

Visual representation matters. When territories are left out, they fall out of mind. The fact that the United States' outline never includes distant territories such as American Samoa or Guam has repercussions on the place of these territories in American consciousness. These unrepresented (and unrepresentable) spaces also challenge the Westphalian ideal of jigsaw-like entities with their homogeneous colors and shapes that neatly fit together, without gaps or overlaps. Pushing issues of cartographic representability even further—and opening a wider breach between the affect-laden logomap and actual management of territorial sovereignty—is another aspect of territorial sovereignty that appears so self-evident that it is hardly ever mentioned. Yet this is a tenet against which new forms of territorial management are also increasingly chafing, and one I discuss below.

Bonus Tenet 4: Immobility

If traditional cartography is poorly equipped to represent nonbordered, nonhomogeneous, and dispersed political entities, an even more significant challenge is to incorporate movement. And it is perhaps in this realm that we more fully grasp the incipient reversal of the role of cartography as technology of sovereign

control. As mentioned in the introduction, while the map initially preceded the territory and anticipated a moment where technology would eventually catch up with it—as witnessed in remote and inaccessible corners of the state—progressive incursions into unmapped realms have overturned this relationship. Now cartography has to devise new ways to represent spaces that are being explored/exploited beyond the horizontal, the enclosed, and the surface. For the last few years, Google Maps has included layers to map out buildings and shopping malls, offering different surface views once a specific floor has been selected. But for more complex spaces such as state territories, representation is more challenging.

In a recent essay, Wayne Chambliss tracks this dialogic evolution between mapping and exploration in the case of the subterranean. State actors, he writes, have been relying on new technologies—such as magnetometers and other metal detectors, electromagnetic induction, electrical resistivity, gravity-measurement technology, and seismic and acoustic sensors—to map the subsurface. Of all these techniques, Chambliss argues, gravity-measurement technology is especially promising. Because Earth's composition is not homogeneous, the force of gravity varies at every point of Earth's surface by a tiny amount. As he explains, the US military is currently mapping the subterranean to create a baseline. Subsequent changes in the gravitational fields, such as the creation of underground bunkers or a military facility, will alter the structure's gravitational signature, making the structure visible. If it is feasible in theory to spoof these mapping technologies (by adding heavy material to counterbalance the missing soil), it will prove increasingly difficult to do so as the technology becomes more sophisticated.[25]

If his research articulates how mapping is already playing catch-up with the mobilization of new territorial entities, other technologies—and particularly those aided by artificial intelligence—are creating an even wider breach between the logomap as established vector of national affect on the one hand, and future forms of territorial management and state contours on the other. Military objectives have increasingly been to find new ways to address state security beyond human geographies and through technologies that would extend—indeed, supersede—human capabilities. In many ways these are not recent developments, as Jeremy Packer and Joshua Reeves show.[26] As early as the 1890s, new forms of technical media, such as photography for instance, were already being mobilized to fine-tune the capacities of artillery and create superior weapons. In its ever-evolving search for technological superiority, the US military is adopting modes of information, communication, and mobility that gradually require less and less human input. In fact, seen as "inherently undependable" and with "highly unreliable data collection, storage, and processing capacities," humans essentially introduce noise and errors in a given system.[27] As a result, foreshadowing what Packer and Reeves term "humanectomy," the US military is experimenting with "taking humans out of as many links in the chain of command as possible."[28]

I have explored elsewhere the common bodily metaphors and analogies that have helped naturalize state outlines and weave affective ties between the citizen and her logomap.[29] However, new organizational models based on biological architectures are privileging very different bodies, such as swarms, that give precedence to autonomy, emergence, and distributed functioning. Comparing favorably with human cognitive intelligence, the swarm requires no planning, central representation, or traditional modeling.[30] As the media theorist Jussi Parikka explains, swarms are organizations that are continuously on the verge of both materialization and dissolution. They are both radically heterogeneous and consistent.[31] While this makes them a superior model to face challenges to territorial sovereignty beyond the horizontal, it makes them a poor vector of patriotic affect. Not only are swarms difficult to represent spatially, but they also feel very alien in their lack of autonomous individuality. The swarm's model is not human but avian, possibly entomological. It also functions in ways that challenge our notions of autonomy and are difficult to comprehend.

There is now a genuine risk that humans may eventually be entirely taken out of geopolitical assemblages. The human component, with its limited vision, its juvenile data-processing capacities, and its highly vulnerable communications processes, is the ultimate source of the fog of war. As sophisticated systems for processing battlefield data are being developed, the human has suddenly emerged as an epistemological hindrance. By contrast, the swarm cloud possesses a continuously refined, emergent collective intelligence that is far beyond the grasp of humans' physiological capacity. Military strategists have now begun to imagine the emergence of a "swarmanoid"—a "heterogeneous assemblage of bots that each perform unique epistemological and kinetic functions."[32] As technology develops further, components of such assemblages are likely to involve scales at the very limit of human perception and detectability. Nanotechnology in particular promises to enter the domain of the microbial—"worlds and microcosms . . . inaccessible to prosthetic-free human experience, zones hard to apprehend as connected to our own forms of life."[33]

At the same time—and this is the core point I want to make in this chapter—this evolution is taking place in parallel to a geopolitical imaginary that remains very much aligned with "traditional" forms of territorial control. While military emphasis on "full spectrum dominance"—an ambition for the total occupation and control of land, sea, outer space, and cyberspace[34]—may suggest as much, this breach is not a neat disjunction between military and civilian imaginaries. The challenge to imagine territorial sovereignty in other ways is one that is shared across publics. Thus the statements by the Russian military theorist Vladimir Slipchenko that the very spatial categories of war are changing, that area-based military concepts such as front, rear, and flank would be irrelevant in the future, and that "there will be no need to occupy enemy territory" appears premature in the current context of the Russian war on Ukraine.[35] While interstate conflicts, and indeed much of neoliberal economies, operate on different spatial logics, military decisions continue to be made on the basis of classical geopolitical imaginaries and logomap-

bound affect. As Martin Coward articulates, traditional notions of modern warfare conceived of a front line rolling across a planar surface, taking possession and control of everything it encompasses. By contrast, in the spatial logics of modern warfare, "points are not related across defined distance but in terms of significance"—they are therefore not spatially fixed, but "shift in dynamic ways according to relationships," creating an entity without edges or distance.[36] But these labile and continually morphing geographies are difficult to accommodate within the parameters set by the four tenets of containment, homogeneity, contiguity, and immobility.

Conclusion: The Rise of the Entomological Subject?

If the gap between map and territory is not novel, what I have suggested in this chapter is that it is widening to the point of having created two irreconcilable spatial imaginaries. There are attempts, as I have shown with the case of Japan's terraqueous map, to integrate into the logomap spaces found beyond human experience, and we can anticipate further attempts at "volumizing" the logomap. Even in the unlikely event these enlargement projects are successful and citizens do develop patriotic sentiments for maritime and aerial terrains, future swarm-like formations and incursions into spaces found beyond human perception or scale ultimately make these attempts futile.

This divergence is also significant in that it signals a rupture between the discrete logomap and the body of the citizen—a powerful analogy that has played a defining role in modeling and structuring national affect. Lacking spatial representations evoking containment, contiguity, and physical integrity, the model of the swarm fails to elicit the same kind of recognition, and the metaphors it elicits are no longer human but entomological. We no longer have a unified and coherent territory, but an ensemble of points—a network continually morphing, disassembling, reassembling.

In a recent book on borders, sovereignty, and security, Matthew Longo spells out the "grand strategic shift *away from a focus on nation-states and sub-state actors and toward the individual*." The point, he writes, is no longer about the *enumerated* subject, but about the *pixelated* subject—one constituted by an aggregation of datapoints.[37] This argument suggests that the evolution of territorial management may not be just a matter of reframing the state's spatial remit, but representing the very dissolution of the body of the citizen as well. This analysis reverberates the concerns of Benjamin Bratton who suggests we view the various types of planetary-scale computation such as smart grids, cloud computing, robotics, mobile software, ubiquitous computing, and the like as forming a large, coherent, but accidental megastructure he calls the Stack—ushering in a new architecture for how we divide up the world into sovereign spaces.[38] These kinds of entanglements across scales, blurring the human/nonhuman, organic/nonorganic, and animate/inert categories, open potential avenues to reframe a Westphalian model, presently predicated on zero-sum calculations, in a more collaborative

way.[39] It would also allow for a better handle on pandemics or meteorological phenomena, which Timothy Morton refers to as hyperobjects. Too vast to be visible in their entirety, because of their transdimensional nature, they often tend to appear nonlocal and temporally foreshortened.[40]

Rather than see enclaves and exclaves, noncontiguous states, free trade corridors, and special economic zones as exceptions, it might be preferable to embrace spatial discontinuities as a more realistic model of territorial sovereignty. Bratton draws attention to the ways the internet is emerging as "a living, quasi-autonomous . . . transterritorial civil society that produces, defends, and demands rights on its own" beyond the body of the state. This does not portend, he insists, the demise of the state, but does imply its ongoing redefinition "in relation to network geographies that it can neither contain nor be contained by."[41] Bratton envisions, for instance, the possible scenario of a quasi-state actor, religious polity, or cloud platform offering more secure levels of digital identity than a state[42]—leading to the coexistence of overlapping realms of sovereignty (national, political, economic, fiscal, etc.) in ways that resemble the pre-Westphalian order.

In such a scenario, attempts at bringing cartographic (and especially logographic) representations in line with the state appear particularly futile. It is perhaps an opportune time to devise alternatives, especially since logomaps are inevitably incomplete (in leaving out islands, territories, and territorial fragments) and misleading (in their use of full colors). Here the case of the United Kingdom might serve as a useful example. In addition to the difficulty of representing the state as a logomap, given that it involves various legal entities both within the state and adjacent to it, such as Crown dependencies (the Isle of Man, Guernsey and Jersey), overseas territories, and Commonwealth realms, the UK's own territorial arrangement of four components—England, Wales, Scotland, and Northern Ireland—across two main islands adds further complexity. Perhaps out of a reluctance to split Ireland, the UK rarely uses its logomap, opting instead for a flag.

Not only does this rupture between territory and sovereignty not appear to have weakened national attachment, but the very nonterritorial representation is in fact more inclusive, since it is no longer making claims to geographical accuracy. We can see similar incipient developments elsewhere—notably, with the French hexagon, represented in more abstract ways in the colors of the flag. Unlike a logomap aligned on an ideal of territorial container, these abstract logos offer membership to a more geographically malleable entity.

In *After the Map*, William Rankin argues that electronic navigation systems are increasingly displacing previous mapping practices, and as a result, the tight relationship between geographic legibility and political authority has been lost. Because electronic coordinates are explicitly designed to exceed the boundaries of individual states, he writes, there is no longer a requirement of a "long-term geographic commitment or sustained control over a contiguous expanse of land."[43] While I fully agree with his thesis that we are witnessing a transformation in the practice of sovereignty, his contention that it has become "possible to imagine territory as something separate from sovereignty" is not supported, in my view, by the global rise in populist ethno-nationalism, nor by the enduring forms of traditional warfare to acquire territory, such as Russia's current war on Ukraine.

Such diagnostics remind me of the forecasts made in the early 1990s by various scholars who saw the collapse of the Soviet Union and the increase in global linkages as harbingers of a new "borderless world." Not only did these predictions never come to pass, but what we witnessed instead was a sharp increase in border walls and a rise in border security. What I have tried to show in this chapter is that the growing complexity of territorial management in three-dimensional space and beyond the core tenets of containment, homogeneity, and contiguity is opening a breach between practice and imaginary and upending the hierarchical and sequential relation between the map and the territory. This particular gap, whose further trajectory I will not attempt to forecast here, deserves attention from cartographers and social scientists. Its evolution is traceable in the new logomaps attempting to encompass spaces beyond the terrestrial (Japan) or, conversely, to move away from the geographical (UK, France). To what extent these efforts will lead to genuine transformations in the imaginary of the state is too early to tell.

PART III

From Critique to Counter-Cartography

8

Barbara E. Mundy

Indigenous Sovereignty Out of Time

Sovereignty, be it of a monarch or of a modern nation-state, is both *de jure* and *de facto*. While it may be articulated through laws, it is also, at the end of the day, a set of practices—often ones that control the movements of bodies and goods. Violent practices may be the means through which sovereign entities exert control, but violence is too rapacious to be sustainable in the long run. As recourse, sovereign powers depend on representations to articulate their claims and to broadcast their authority.

The cartographic form has proved to be a robust vehicle for sovereign powers, making visible the territorial dimensions that serve as a yardstick of the state's spatial expanse. Despite the frequency of cartography's use to document the sovereignty of modern nation-states, however, there is nothing particularly natural about this association, and in other places and other times, sovereignty over a particular territory has been imagined and expressed in a myriad of different ways. For instance, in the empire created by the Aztecs (Mexica), political sovereignty over a territorial expanse was expressed through the architecture in their capital city, Tenochtitlan; the *huei tzompantli*, the great skull rack, neatly displayed the skulls of captured warriors from conquered pol-

ities to express Mexica sovereignty over distant territories and their peoples, akin to the Roman display of the spoils of war in their imperial capital to symbolize their control of distant lands.[1] Given that the relationship between the territorial map and political sovereignty is a fairly recent one, at least within Europe, what are the historical origins of this linkage? And since such maps are necessarily silent about competitive claims on sovereign space, how might they be reinterpreted to reveal alternate forms of territorial imaginaries?

For me, as an art historian of Latin America with a focus on the art and cartography created by Indigenous communities of Mexico, the political stakes of these questions are clear—because in the Americas, the claims by European powers of their sovereign right and coincident territorial possession were always contingent on Indigenous *dispossession*. Following the arrival of Europeans in the Americas, beginning in the fifteenth century, empires like the Aztec and the Inca saw their leadership decapitated, their accumulated riches seized or squandered, their rights to land ignored, and their political authority dismantled. Other scholars have shown that the map played a role in dispossession. What I will argue in this essay is that the cartographic history of Indigenous dispossession depended on two widespread tropes. The first was the cementing of the association between territorial map and sovereignty in the early sixteenth century, which had as its aim and consequence the displacement of Indigenous peoples as sovereign rulers.[2] The result of this bond is patently clear in nineteenth-century maps of the United States, where changing representations of national territory showed the acquisition and annexation of lands to the west and south of the Atlantic seaboard core. Left off these maps altogether were Indigenous peoples and their once-sovereign claims over territories and resources—an absence that Jen West, in ~~Ohlone~~ *Stanford Lands* (discussed in the introduction and reproduced as figure 0.1), marked via its erasure. In Latin America, equivalent cartographic projects were undertaken to make visible new nations that emerged in the nineteenth century.[3] But the dominant presence of Indigenous populations, particularly in Mexico, made it impossible to leave them off the map. Instead, mapmakers drew on another trope: setting Indigenous peoples, and their claims to sovereignty, *out of time*. By this I mean, following Johannes Fabian, that Indigenous people were denied coevalness with the viewer.[4]

The form of this essay is admittedly temporally peripatetic, but it begins and ends by considering the afterlives of the same Aztec map. Moreover, all the sections are anchored in close examination of individual maps—most of them published ones; most showing Mexico, if not urban Tenochtitlan; most of wide circulation and impact. As cultural ar-

tifacts, these are always open to new interpretations over time. The first section begins in the nineteenth century with a map of the nation-state to show how the "problem" of Indigenous territoriality was a live issue within the non-Indigenous sovereign state (in this case, the nation-state of Mexico) still unresolved after centuries of European expansion. To explore the origins of the cartographic history of Indigenous dispossession, it turns back to the sixteenth century, to look at one of the earliest and most famous maps to depict an Indigenous city, that of Aztec Tenochtitlan, which was published in the German city of Nuremberg in 1524. This map is central to my argument that the sovereign claims made via territorial maps hinged on Indigenous dispossession, and it serves as a lodestar through the rest of the essay. In subsequent sections, I follow later manifestations and editions of this map of Tenochtitlan, to show something of a history of its reception and ongoing reinterpretation, as well as to trace the rise of a new trope, that of Indigenous people as out of time.

Since all maps are, at the end of the day, imaginative projections staking a claim to truth, it is worth taking the work of imagination, and its relation to truth making, seriously. Thus, the final section of the essay turns to the work of two contemporary artists, Sandy Rodriguez and Mariana Castillo Deball, whose works interrogate the relationship between cartography, imagination, and claims to truth. Both deploy the powerful language of cartography to imagine new projections of Indigenous territoriality, both in Mexico and within the United States, with Castillo Deball using the 1524 map of Aztec Tenochtitlan as a springboard for a number of her pieces. In their complex and multifaceted works, these two artists offer a counter to the map as a vehicle of Indigenous dispossession, and a corrective to the long history of depicting Indigenous peoples of the Americas as "out of time."

Nineteenth-Century Nationalist Mapping

A well-known atlas, *Atlas pintoresco e historico de los Estados Unidos Mexicanos* (1885), reveals the complexity of setting Indigenous peoples out of time in the Mexican context.[5] Created by one of Mexico's best-known geographers, Antonio García Cubas (1832–1912), the atlas showcases the national territory of Mexico. Each of its large and beautifully illustrated folios has at its center the familiar cartographic image of Mexico, its shoreline set against the waters of the Atlantic to the east and south, and the Gulf of Mexico to the west. Above its northern border, the lands of the United States are a muted brown. Most Mexicans looking at the map at the time of its production would have recalled the reduction brought about by the humiliating Treaty of Guadalupe Hidalgo of 1848, which ceded over half of Mexico's territory to the United States. As Raymond Craib has argued, after this loss, "[Mexican] national maps acquired even more importance as a means to set the bounds of territorial sovereignty and to provide a textual tangibility to an otherwise metaphysical entity."[6] A close analysis reveals that these maps also present competing claims on national space. As the reader turns the pages of the *Atlas pintoresco*, the national outline repeats, but the interior forms, adjacent statistics, and framing elements change. Each page presents the nation as seen under different lights.

The first map of the atlas is a "Carta Politica." Showing the national territory as divided into component states, it is framed by pictures of government buildings and portraits of Mexican leaders since independence from Spain in 1821. The emphasis of this opening map is the men and infrastructure that modern governance comprised. Deeper into the atlas, a *Carta Etnografica* shows a very different national territory (figure 8.1). Here, García Cubas divides the national map into zones of different colors, each one marking the territory of Indigenous groups, as defined by language and customs. Related census statistics are shown in inset graphs. On the bottom left, *población en general* (general population) reveals *mestizos* (people of mixed European and Indigenous heritage) to be the dominant group, followed closely by *raza indigena* (Indigenous races). By contrast, *raza blanca* (white race) is less than half that figure.

The map celebrates the nation. It also reveals the weakness of the Mexican national project over the territories within its borders given that the Indigenous communities shown on the map maintained an autonomous capacity for world-making, particularly through the use of Indigenous languages and customs such as pilgrimages to sacred sites, which allowed them to imagine and express connections to the territories they claimed as their own.[7] In all this, Indigenous practices pushed back against the modern sovereign state, most of all its language policies. Since the sixteenth century, it has tried to get its citi-

zens to adopt a single language that would "know everything, say everything, and translate everything," to establish its hegemony.[8] In Mexico, the project to impose the Spanish tongue gained traction during the late eighteenth century through educational reforms promoting Spanish literacy and language use in Indigenous communities.[9] So when García Cubas included this map in his atlas showing the rich plethora of Indigenous languages, he was unwittingly marking the failure of the state to assert its linguistic (Spanish-language) sovereignty.

At the same time, by depicting the number and varieties of linguistic groups within Mexico, the map reveals a deep history of Indigenous presence on the land and territorial claims to it: languages and customs do not sprout overnight, like mushrooms in the grass. They are the human equivalent of a fossil record, revealing a human deep time. In the case of Mexico, the long-standing Indigenous presence on the territories is likewise registered in the abundance of native-language place names that can be seen on the maps included in the *Atlas*, including "Mexico"

FIGURE 8.1. Antonio García Cubas, *Carta Etnografica*, in *Atlas pintoresco e historico de los Estados Unidos Mexicanos* (Debray Sucesores, 1885). Public domain. Courtesy of the Library of Congress, Geography and Map Division, Washington, DC.

itself. For García Cubas and other intellectuals of his era, such native names were a double-edged sword. As Craib has argued, the names were a "living source of [a] deep history" that distinguished Mexico from its shallowly rooted neighbor to the north.[10] At the same time, the colored zones on the map also signaled resistance to the national language of Spanish. For optimistic educational reformers of the nineteenth century, Indigenous ignorance of the national language was a problem the future progress of the nation would resolve.

A Cartographic History of Indigenous Sovereignty

The "problem" of an Indigenous territoriality that was resistant to the sovereignty of the nation-state, detectable in the *Atlas pintoresco*, has a long cartographic history, with its modern origins in the late fifteenth and early sixteenth centuries. As we will see, the map became one of the public stages on which European monarchies acted out their imperialist missions, which included bringing millions of once-autonomous Americans under their sway. This was particularly true of the Habsburgs, who were consolidating political sovereignty over scattered and contentious states. The conquest of the Aztec Empire gave Spain's Habsburg ruler an unprecedented, and unexpected, territorial reach. Necessary to royal consolidation of power was the dispossession of Indigenous monarchs in the Americas—which in turn required undermining their sovereign claims. A "polemics of possession" thus marked the chronicles,

histories, and public debates that raged across the sixteenth century.[11]

One beginning point for the erasure of Indigenous sovereignty might be 1493, when Pope Alexander VI granted all the lands lying a hundred leagues west of the Azores and Cape Verde islands to the Catholic monarchs of Spain and Portugal, a land giveaway that was later modified in the 1494 Treaty of Tordesillas. Jordan Branch (whose work is included in this volume, chapter 6) has argued that the treaty allowed Spain and Portugal to "make political claims based on cartographic territoriality."[12] Yet despite the fact that the treaty coincided with a period of growing technological sophistication in nautical cartography, as well as developments in printing that allowed for wide dissemination of maps, no map was produced as part of the agreement to document or visualize these sovereign claims. Instead, it was about twenty-five years later that the process of cartographic erasure of Indigenous sovereignty began.

The watershed moment comes with the 1524 publication of a map of the great Mexica (or Aztec) capital of Tenochtitlan (figure 8.2), to accompany the text describing its "discovery" by Spanish conquistadors. The woodcut map presents the city, today known as Mexico City, as it appeared around 1520, when it was under the control of Moteuczoma II. Historians estimate that the region may have been home to as many as three hundred thousand people, equal to Europe's largest cities at the time.[13] At the center of the map is the island city in which Moteuczoma held sway; around the lake are other cities shown at smaller scale. Along the edge there appears a separate sche-

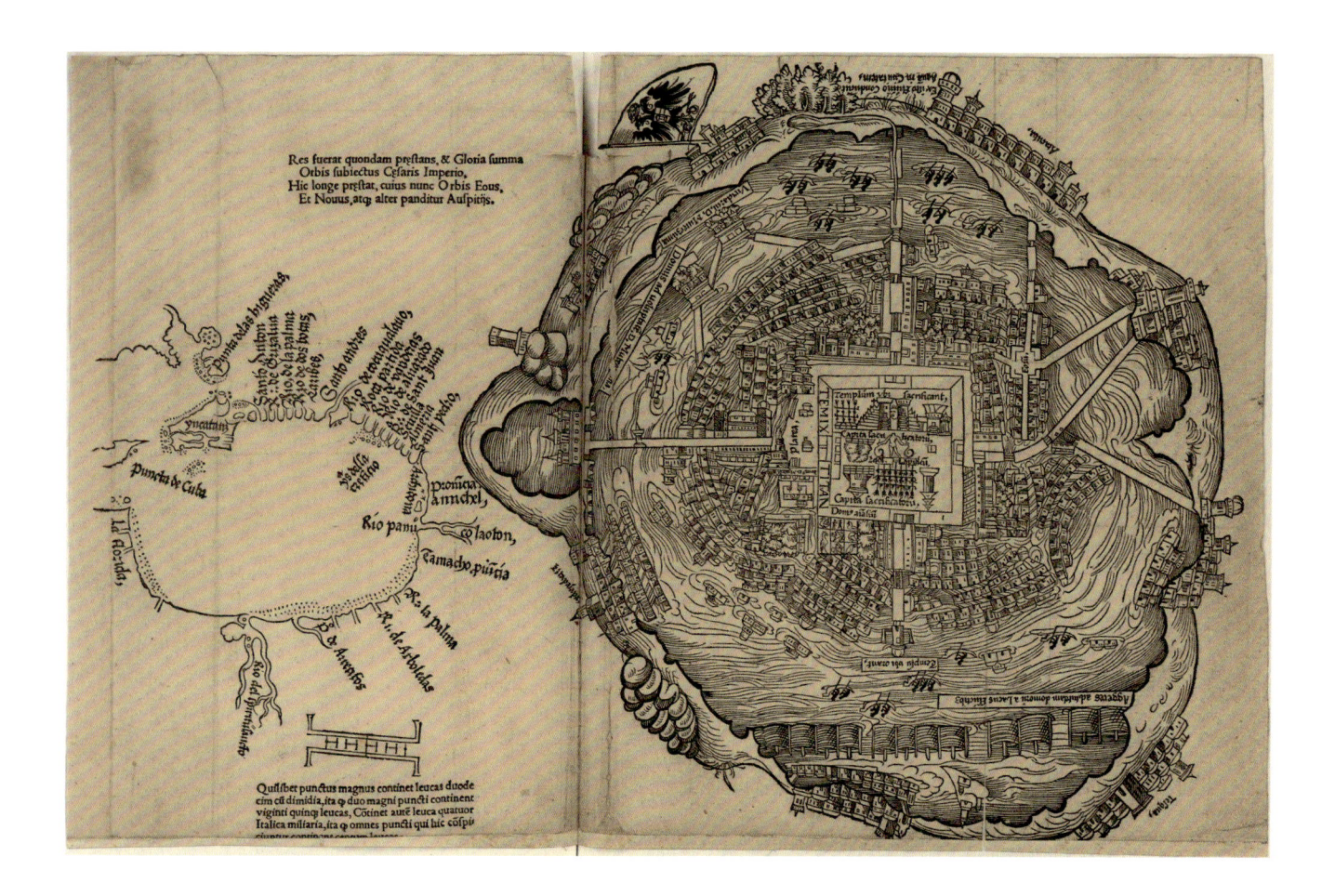

matic map of the Gulf Coast. This was where Spanish forces, under the command of Hernán Cortés, had landed the previous year.

The maker of this woodcut map chose to make the city its central focus, and this corresponded to the political realities in both the Americas and Iberia, for it was within cities that sovereignty was located. In Spain, *cabildos* (Spanish: town councils) held considerable amounts of political power, as did the *altepetl* (Nahuatl: city-state) in the Americas. Tenochtitlan, the seat of Moteuczoma, itself has been described as a "complex altepetl" because of its size and structure of governance.[14] Even more important is that on both sides of the Atlantic, cities were equated with advanced civilization. The Mexica had poured resources and human capital into their

since its official founding in 1325, and it was within the urban forms of the newly encountered continents that Spanish observers found powerful evidence of Amerindian civility.

Through text and image, the creator of the woodcut map marks Moteuczoma's presence by naming his palace, his gardens, and the "pleasure houses" that housed his wives. His palaces stand adjacent to the great Sacred Precinct, a square zone in the center of the city; here appear the great Main Temple, the stepped twin pyramid, a massive sculpture, and the *huei tzompantli* (although its constituent skulls are indiscernible as such) along with other temple structures. We can also read the inclusion of the Gulf Coast map at left as another index of Moteuczoma, as it reveals the eastern reach of his domain. Read in this

FIGURE 8.2. *Map of Tenochtitlan and the Gulf of Mexico,* in Hernán Cortés, *Second Letter [to Charles V of Spain]* (Nuremberg, 1524). Public domain. Courtesy of the John Carter Brown Library, Providence, Rhode Island.

light, the map presents the Mexica capital as part of the secure domain of an Indigenous monarch, with an urban order that was at that moment a European ideal.

But one aspect of the map has barely been noticed: it was visibly amended to assert Charles V's sovereignty over this territory, not that of Moteuczoma. The Mexica prototype of the map likely showed the city as it looked in 1520, before the assassination of Moteuczoma in June of that year. I have argued elsewhere that the map was based on one made by an Indigenous creator, this prototype making its way to Europe around 1520, likely to accompany a letter Cortés had sent to Charles V, known as the Second Letter.[15] This was hardly a triumphant missive; while in it Cortés describes the marvels of Tenochtitlan, he wrote it right after his forces and those of his native allies had been expelled from the city by Mexica warriors, and he was in retreat. When the Second Letter was first published in Seville in 1522, it lacked a map. Nonetheless, it was a sensational hit, and two years later, a Latin edition, along with the map, was being prepared for publication in Nuremberg. But by then, news of dramatic events of 1521 had reached Europe. After a brutal siege of the city by Cortés and his Indigenous allies, the Mexica had capitulated. The 1524 map, first rendered to show the sovereign domain of an Indigenous monarch, was hastily amended, with a text that alluded to breaking news. It reads, in Latin,

Res fuerat quondam prestans, & Gloria summa
Orbis subiectus Cesaris Imperio

Hic longe prestat, cuius nunc Orbis Eous,
Et Novus, atq alter panditur Auspitiis

(This was once the foremost thing. Even the
   highest glory
in the world, after becoming subject to Caesar's
   empire/reign,
thus improves by far—he whose world now to
   the East,
and the New, as well as the old, extends under
   his auspices.)[16]

The "thing" of the first line refers to either Tenochtitlan or the territorial expanse on the map as a whole, while "once the foremost" refers to its position within the domain of the Mexica, as the most powerful city. These words describe the city as it may have been in 1520. Its conquest follows. In the second line, the "Caesar" to whom Tenochtitlan is now subject is none other than Charles V, the recipient of the letter. Charles was crowned king of Spain in 1516, but by 1519, his realm was extended eastward when he was elected Holy Roman Emperor, as alluded to in the third line. The last line of the text contrasts his eastern empire to what is shown on the map, the "New World" now under the "auspices" of his sovereign power. Working in a print shop in Nuremberg, a city that was under Habsburg control, mapmaker and author of the textual addition alike show their eagerness to flatter the powerful king. In a visual echo of the text, the mapmaker has planted the Habsburg banner at the center line and top of the map. Thus this map, first designed to show, with no small admiration, an autonomous Indigenous city, was transformed into a token of its conquest.[17]

The "improvement" that the city was to enjoy by becoming part of Charles's domain was confident rhetoric that masks a great deal of political uncertainty. Procedurally, that Cortés should unseat an Indigenous monarch created a problem for the conquistadors and their ruler, Charles V. The legal codes of the day proscribed that one monarch should seize, without reason, the lands and peoples of another "natural lord."[18] Cortés, who had some legal training, was seemingly aware of this potential violation. Already in the carefully crafted text of the Second Letter he began to spin out a justifying scenario that would have consequences for the meaning of the map as a token of sovereignty. In describing one of his encounters with Moteuczoma, Cortés claimed that the Mexica monarch delivered a speech to him, in which he acquiesced to accept Charles V as his overlord. Despite speaking no Nahuatl, Cortés set Moteuczoma's words down as if transcribed on the spot:

> According to the direction from which you say you have come, namely, the quarter where the sun rises, and from what you say of the great lord or king, who sent you hither, we believe and are assured that he is our natural sovereign, especially as you say that it is a long time since you first had knowledge of us. Therefore be assured that we will obey you, and acknowledge you, for our sovereign in place of the great lord whom you mention, and that there shall be no default or deception on our part.[19]

Setting aside the dubious facticity of this speech, its rhetorical value was immeasurable, neatly answering the question of whether Charles V actually had any rights over another sovereign emperor and his kingdoms: it purports that Moteuczoma himself pacifically accepted Charles as overlord.[20] Also crucial along with this supposed transfer of sovereignty was that Moteuczoma made Cortés the gift of a number of maps, one of which may have been the prototype for the map published in Nuremberg in 1524.[21] In this context, the proffered Mexica maps served as a visible token of Moteuczoma's territorial possession. When passed from the Mexica monarch to Cortés and then sent on to Charles, they concretized the promise of Moteuczoma's speech and the transfer of sovereignty. Charles V, as the new and rightful "natural lord," now held the map. With its publication—which included the text quoted above setting Tenochtitlan as part of Charles's *Cesaris Imperio*—engagement between map and royal sovereignty was fleshed out for the first time to a European public.[22]

Maps and Embodied Sovereignty: An Open Question

The 1524 map crystalized the territorial map as token of sovereignty, both in the form of a gift from Moteuczoma to Cortés and in the map itself, whose text makes clear that this *res*, or "thing" (the paper map of territory), is meant to represent the "extension" of Charles V's dominions into the "New" world. Its clarity is all the more striking in that it was created in a period when the sovereign claims of many European monarchs (Charles V, Henry VIII, Francis I) were tentative. Royal domains were often scattered, and monarchical power was patchy. Moreover, the rights of "natural lords" were still under debate. Concurrently, the visualization of royal sovereignty was in an experimental phase as well. At least one cartographer, Johannes Putsch, drew on medieval theories of the king's two bodies as the intellectual backdrop for his map: one of these bodies was mortal, the other transcendent, containing divinely invested sovereign rights.[23]

This notion of the conceptual and enduring body of the monarch was a useful conceit, allowing mapmakers to combine figuration with cartography. Putsch's 1537 map, *Europa regina*, was a tentative attempt in this direction.[24] In a map made fifty years later, based on Putsch's design (figure 8.3), all of Christian Europe is imagined as a woman: a striking image rooted in what Elke Anna Werner has described as "the powerful tradition found in Medieval political theory of the unity of all Christians."[25] Spain wears the crown and Bohemia is the heart, reflective of Habsburg control of Europe entering its peak.

These radical experiments in human cartographic figuration did not endure into the seventeenth century, in no small part because of an inherent tension introduced when figuration, one mode of representation, is introduced into cartography, an entirely different mode.[26] The resulting clash of symbolic systems, Christian Jacob points out, "[turns] the space they represent into an impossibility, a geographical paradox."[27] Landscape views, which presented a place as it might be seen by the human eye,

resolved some of the visual tension by combining abstracted landmasses (rendered at one scale) with mimetically represented human bodies (rendered at another). In the landscape views that were created across the sixteenth century and beyond, the foreground of the map often served as a stage on which to display the particular peoples who inhabited that space, often in male-female pairs. These figures were far more than decorative. As Valerie Traub has argued, they reveal "an increasing tendency to view bodies as potentially classifiable objects of an epistemology organized according to ethnographic and gendered imperatives."[28]

In the case of maps of Mexico City (as Tenochtitlan became known during the sixteenth century), ethnographic "types" appear, but the epistemology was marked by temporal uncertainty. The 1524 map itself had a long afterlife, like a comet with a long tail, because it was reworked across the sixteenth century, sometimes to accompany a new edition of Cortés's letters, sometimes in a different context. One of its best-known instances is within the first extensive atlas of city maps and views published in Europe, *Civitates Orbis Terrarum*. Here, it was part of a bifolio, with Cuzco, the great Inca capital city, on the facing page (figure 8.4). To represent Mexico City, the artist added to the 1524 map and flipped it along the vertical axis, resulting in a reversed image. The city retains many of the architectural features found in the prototype, including the Sacred Precinct with its twin temples and a massive figural sculpture, but their names are erased. No longer is Moteuczoma encountered on the map's surface. Instead, in the foreground, eth-

FIGURE 8.3. Matthias Quad, *Evropae Descriptio*, Cologne, 1587. Staatliche Bibliothek Regensburg. Public domain. Available at https://daten.digitale-sammlungen.de/~db/0011/bsb00113022/images/index.html. Reproduced under a CC BY-NC-SA 4.0 DEED (Attribution-NonCommercial-ShareAlike 4.0 International) License.

FIGURE 8.4. Georg Braun (1541–1622) and Frans Hogenberg (1539–1590), *Mexico and Cuzco*, in *Civitates Orbis Terrarum* (first pub. Cologne, 1572). Public domain. Courtesy of the David Rumsey Historical Map Collection, David Rumsey Map Center, Stanford Libraries.

nographic figures are set into the stage-like space; one of them, who stands at center wearing a tunic and feathered headdress, is attended to by a warrior at left and what seems to be a female companion at right. The suggestion of high status, through garb and bodily composure, is consistent with other such figures in the atlas.

Crucially, the temporal status of these figures is ambiguous, as is that of their cities. The map is labeled "Mexico, Regia et Celebris, Hispanae, Novae Civitas." When it was created, the city was the capital city of the Viceroyalties of New Spain. It boasted a majority Indigenous population, under the sover-eign control of Spain's Habsburg monarchs. Looking closely at the map, we find no figures dressed in European clothing, even though many prominent residents of the city would have dressed in this manner, including Indigenous elites. Moreover, the rendering of Mexico City preserves its pre-Hispanic architecture, even though that architecture was mostly destroyed in the 1520s when Spaniards occupied the city. For the historically informed viewer today, the dissonance between the moment the map was created and the moment it presents is heightened by the claims of the atlas in which it is found. That atlas's creators, Georg Braun and Frans Hogenberg, boasted of presenting *contemporary* views of cities, as suggested by the title on the map itself.

To some extent, the anachronic presentation of Mexico City may be attributed to the tight control that Habsburg rulers exercised over the print markets within their spheres, refusing the necessary licenses to publish works about their American possessions that they deemed prejudicial to their image or interests.[29] European travel to the Americas was likewise constrained, so even had they wanted, the Cologne-based Braun and Hogenberg would have had a hard time accessing up-to-date views. As a result, the 1524 map was largely recycled for *Civitates*—with just enough changes, including the addition of figures in the foreground, to give it a semblance of the "new." Responding to opposed forces—the public's desire for something new, and the Habsburg desire to censor unfavorable news—mapmakers' responses left the figures of Amerindians, and the cities they inhabited, in a temporal limbo. The image of the Amerindian was drawn from outmoded sources but given a veneer of contemporaneity. Given the lack of direct contact with Amerindians and the scarcity of news about them, most Europeans encountering such time-warped images would have been none the wiser. It was thus that the image of the timeless Amerindian lived on in Europe.

Transformation of the 1524 Map

Across the sixteenth and seventeenth centuries, the 1524 map continued to be reworked, even as it grew further removed from the actuality of the city it rep-

resented. During the course of revision, it was flipped and altered but retained key identifiable features, like the buildings of the Sacred Precinct. The change effected by Braun and Hogenberg's atlas, where the city appears in a somewhat oblique bird's eye view, was visually harmonized into a fully oblique view, with the same stage-set foreground. One of the most important versions of this type was created in 1671 for John Ogilby (1600–1676) and Arnoldus Montanus's (1625?–1683) *America: Being the Latest, and Most Accurate Description of the New World*. Despite the promise of up-to-date information, the full-folio map of Mexico City in this work presents it at some past moment, as indicated by its title, "Vetus Mexico" (Old Mexico) (figure 8.5).

Temporal ambiguity is introduced in the foreground of the map, where the mapmaker set figures in European dress of the time—the hatted and caped white figures in the right foreground—as well as familiar transport technologies, like the carts pulled by oxen in the background. The multiracial nature of the city, a legacy of African enslavement, is cued by the dark-skinned figures who labor under heavy packs or rest from their exertions. The pages around the map highlight Mexico's past, not its present, including stereotypical "Indian" figures wearing feathered skirts and headdresses. The point is not just that Mexico's Indigenous past was of interest to European readers, but that even in the places where one might expect to find some discussion of Mexico's present, that present did not exist.

Publications like Ogilby and Montanus's had wide circulation, particularly among elite European

FIGURE 8.5. John Ogilby (1600–1676) and Arnoldus Montanus (1625?–1683), *Vetus Mexico*, from *America: Being the Latest, and Most Accurate Description of the New World . . .* (John Ogilby, 1671). Public domain. Courtesy of the David Rumsey Historical Map Collection, David Rumsey Map Center, Stanford Libraries.

readers. The number of translations and editions over the years provides one index of how much Mexico City continued to fascinate European publics. Harder to glimpse is the way such images were received, beyond patchy data about publication (we rarely know print runs). But two years ago, a curator at the Lewis Walpole Library at Yale contacted me about an unusual item in their collection that related to the 1524 map (figure 8.6). It was occasioned by the British craze for "extra-illustration," a leisure-time

pursuit among an educated and affluent class. As I learned, a creator would personalize a printed book by unbinding it, then interleaving the text with hand-selected images that illustrated passages, places, or people mentioned in the text. When the whole collation was rebound, the illustrated material could transform a modest, one-volume book into a unique multivolume edition.

In this case, an unknown creator chose a nineteenth-century book—L. B. Seeley's *Horace Wal-*

FIGURE 8.6. *Mexico*, from M. C. D. Borden's extensively extra-illustrated copy of L. B. Seeley, ed., *Horace Walpole and His World: Select Passages from His Letters* (Seeley, Jackson, and Halliday, 1884), 6:134. A creator whose name is currently unknown chose this nineteenth-century book, unbound its pages, and then rebound them with selected images from different sources, mainly engravings or illustrations published during the eighteenth and nineteenth centuries. Public domain. Courtesy of the Lewis Walpole Library, Yale University.

pole and His World: Select Passages from His Letters (1884)—unbound its pages, and then rebound them with selected images from different sources, mainly engravings or illustrations published during the eighteenth and nineteenth centuries. The amount of new material ballooned Seeley's one-volume text into a twelve-volume illustrated set. Most of the inserted images are portraits: if the text referred to George III, the book's owner would hunt down a picture of George III. But in the case of figure 8.6, the text cites

a passing reference to Mexico that the British writer Horace Walpole once made.[30] That passage clearly excited the reader's imagination, inspiring him or her to locate a printed map of Mexico City.

This map is clearly derived from the 1671 "Vetus Mexico," but certain details (including the fact that it is etched, not engraved) show that it was not directly taken from that source. Wanting to avoid excessive bulk of a paste-over, the creator circled the desired portion of the map in reddish chalk and cut it out. She or he then cut out a circle of a slightly smaller dimension from a plain sheet of paper to create a blank frame around the etched map. The etched map, now framed with blank paper, was next pasted down on another leaf of paper that served as backing. The creator then filled this surrounding frame with a hand-painted scene of imagined inhabitants of Mexico City, a scene intended to seamlessly extend the map's imagery. In adding two prominent figures at left, the creator emphasized the city's past. This "history" was based less on historical sources than on a European imaginary of the Aztecs before contact. Most prominent in the scene is a proud warrior, who stands looking out from the scene. To his right stands a figure carrying what appears to be a handbag containing a pineapple. Both stand on real ground, and they seem to be arrested in the middle of a conversation. In other words, they are not allegories. At the same time, they are out of synch with the city beyond them. There, the pre-Hispanic architecture has been erased, and European rounded buildings guard the entrance to the city. By juxtaposing contemporary Mexico City with images of Aztec warriors from centuries before, the creator of this pastiche effectively placed Amerindians out of time.

While this image could be dismissed as a singular example made for private pleasure, it sits squarely within a longer history of the trope of Amerindians out of time. In selection and presentation, the nameless creator offers a rare example of how the 1524 map was received, and what kind of imaginative response it triggered—a history of reception. And while this

may be a one-off, it is also the considered response of a cultured and educated person, living half the world away. We do not know the creator's identity, but anyone living in Britain in the late nineteenth century would be likely to perceive him- or herself as sitting at the center of a newly expansive empire, holding sway over distant, dark-skinned peoples.

Perhaps as important is that the deployment of this collage—which must have been inserted into the book sometime after 1884, when the volume it illustrates was published—coincides with the codification of cultural anthropology as a discipline. In Britain, the most influential anthropologist was Sir Edward Burnett Tylor (1832–1917), who wrote *Primitive Culture* (1871). Tylor's evolutionary view of human culture maintained that all civilizations pass through stages, from "savage tribe" up to modern complex societies, and this progression had certain milestones—the use of metal, for instance, marked a stage after the use of stone tools. Tylor developed many of these ideas in an account of his trip to Mexico, published as *Anahuac: Or Mexico and the Mexicans, Ancient and Modern* (1861). It reveals his pejorative view of Mexican Indians, who he believed had arrived from Europe long ago; their use of stone tools in the pre-Hispanic period, rather than metal, confirmed that their origins had been as a "barbarous, nomadic tribe."[31] In his travelogue, he sometimes referred to the "Indians" that he met as "Aztecs." In Tylor's schema, Mexico's Indigenous people stood at an earlier moment of cultural development, as if stuck in the past, in contrast to his own putatively developed Britain. His use of the descriptor "primitive," in the sense of "early stage of development," definitively set Mexicans out of time.

Indigenous Sovereignty Out of Time— and Back in Synch

As this essay has traced in broad strokes, the 1524 map of Tenochtitlan was fundamental in cementing an association between cartography and sovereignty, where Moteuczoma's "gift" of a map was presented as the physical token of his voluntary abdication in favor of Charles V. That same act of abdication appears in the speech Cortés attributed to him. (This historiographic erasure of Indigenous sovereignty by a colonizing power finds a parallel in the Mongol case described by Lhamsuren Munkh-Erdene's essay in this volume, chapter 2). In subsequent versions of the map, Moteuczoma's name as possessor was effaced, and instead generic figures of Indigenous people were introduced into the space of the map. But the European fascination with the American past, coupled with a lack of information about its contemporary realities, led to recycled images of "Aztecs" from the past as part of the standard cartographic representation of Mexico well into the nineteenth century, as seen in the extra-illustrated volume. These coincided with, and perhaps contributed to, an emergent anthropological view of Amerindians as people out of time. As seen in extra-illustrated painting, these Amerindians are always depicted in an indeterminate past.

Such tropes have proved enduring. In North America, we encounter versions of this even today with tropes of dead or dying Indians. I hardly need dwell on how pernicious such tropes were, and continue to be, in the effacement of Indigenous sovereignty in the United States. In giving this chapter the title "Indigenous Sovereignty Out of Time," I intended for a certain ambiguity. One meaning is clear from the discussion above: figurations of Indigenous bodies locate them in some ambiguous temporal space with their outlandish feather costumes and antiquated weapons. Such figuration serves to make Amerindians simultaneously real yet not of this moment. But the title could have another reading, akin to Cassandra's warning, that posits Indigenous sovereignty as literally running out of time. This is vividly seen across the Americas. In the US, the lands of purportedly "sovereign" Indian nations have long been exploited with little benefit for their sovereign owners; in Canada, legal disregard of First Nations' sovereign claims have led to massive protests recently over the Coastal GasLink pipeline; in Brazil, the territorial sovereignty of Amerindians in the Amazon is fast eroding as the forest is opened up for ranching and mining. The resultant destruction of the Amazon rain forest is increasingly seen by climate scientists as one of the horsemen of an arriving climate apocalypse.

How might maps bring Indigenous sovereignty back into time, asserting the coeval presence of Amerindians? That is, how might maps be used to destabilize the seemingly inviolate sovereign claims of the nation-state made through cartography, which is the inheritor of earlier modes of sovereignty? Two contemporary artists, Sandy Rodriguez and Mariana Castillo Deball, offer provocative ways to think about these questions.

Sandy Rodriguez lives and works in LA. As a Chicana artist, she has long been interested in representing the history of dispossession of Mexicans and Mexican Americans, today and in the past.[32] Rodriguez was for many years an educator at the Getty Museum, which has a digital project to translate one book of the great Florentine Codex (a sixteenth-century book that has been aptly described as "an encyclopedia of Aztec life"). The bilingual text (Spanish–Nahuatl) of the Florentine Codex was richly illustrated by Indigenous artists, whose images reflect the Aztec past as well as report on contemporary events, like the war of the Spanish invasion of 1519 to 1521.

Rodriguez draws on these arresting images. A recent work of hers—one page of the suggestively titled *Codex Rodriguez-Mondragón*—fills in the cartographic outline of California with images from the Florentine Codex, as well as images of her own invention. Along the Pacific coast, the marvelous ocelot fish recorded by sixteenth-century Aztec artists now swim, as natural histories of the Americas created by Indigenous peoples of the past are transplanted to the present. In her collage, the US–Mexico border—so essential in the current moment in discussions of US sovereignty—has been erased. Instead, images of the desert landscape fill the space, as if to make the point that in the natural world, to which we all belong, such frontiers are fictional.

In another work, meant to protest the detention camps along the border, Rodriguez mapped the borders but undercut the "indigenous out of time" convention (figure 8.7). Instead of showing Indigenous bodies placed in some ambiguous temporal space,

she drew images of Aztec women from the Florentine Codex and set them, weeping and crying, alongside the tents of the detention camps, as if people of the past were actively bearing witness to events of the present (figure 8.8). In this work, the international US–Mexico border serves as a road for the indigenous peccary, also drawn from the Florentine Codex. While Rodriguez's media—paper, ink, and pigment—seem conventional at first glance, they actually derive from a long history of Indigenous image making. For instance, Rodriguez often paints on fig bark or amate paper, the traditional paper used by the Aztecs; she likewise produces many of her own pigments, often by foraging in local semiarid environments, or uses traditional pigments from Central Mexico.

Sandy Rodriguez draws on the communicative power of the national map but uses figuration to bring Amerindians—be they weeping women from the Florentine Codex, or contemporary Mexicans of Indigenous descent—into the same cartographic space. In her abundant images of the natural world that exceed borders, she also points to transnational and transborder phenomena that we, as human animals, are part of.

Mariana Castillo Deball was born in Mexico City; she currently lives in Berlin. I came to know her because of her engagement with historic maps.[33] In 2013, she was asked to do a large installation in an art space in Berlin, for which she designed her *Nuremberg Map of Tenochtitlan* (figure 8.9). The work has several components. The floor has been covered with sheets of plywood, onto which is engraved the

FIGURE 8.7. Sandy Rodriguez, *De los Child Detention Centers, Family Separations, and Other Atrocities*, 2018, hand-processed dyes and watercolor from native plants and earth pigments on amate paper, 47 × 94½ in. Courtesy of the Mellon Art Collection.

FIGURE 8.8. Detail from figure 8.7.

1524 map of Tenochtitlan. Poles have been set up in the space, and upon them are set dance costumes. Other poles hold recreations of symbols that Castillo Deball discovered in ancient Mexican books. Why was she drawn to the 1524 map, I asked her? She responded that in Mexico, this map is the kind of icon one always sees in textbooks, or in history exhibitions. Since the map was made in Nuremberg, Germany, it seemed a natural starting point for an exhibition to open in Berlin.

In order to make the map floor, Castillo used a technology that is as fresh and new in our moment as the printed woodcut was in 1524: she made a digital scan of the image and fed the data into a CNC (computer numerical control) machine, which in turn engraved the lines of the map, at a much-aggrandized scale, onto the planks of plywood. At this step of the process, the map image was difficult to see, as it was simply a network of grooves in wood. Revisiting the same technique that was used to make the wood-

cut image in 1524, Castillo then inked the plywood boards and pulled prints from them.

A photograph taken in her workshop reveals something of the process of creation (figure 8.10). At center, Castillo's assistant, Anna Szaflarski, inks one of the large wooden panels, using a roller to apply the greasy ink in an even layer. In the upper left, another studio assistant, Ayami Awazuhara, gently picks up a sheet of dampened paper from a stack—one pair of clean hands and one pair of dirty hands being the rule of the print shop. She will bring it over to lay on the inked plate to create a print, later to be bound in an atlas. The drying sheets from earlier iterations of this same process are seen hanging like clothes on a line, at the right of the photo. Each print is only a small portion of the entire map image, and is printed in reverse—that is, black background and white lines instead of the original of black-inked lines. Those reversals, of course, echo the long history of print-induced reversals of the 1524 map.

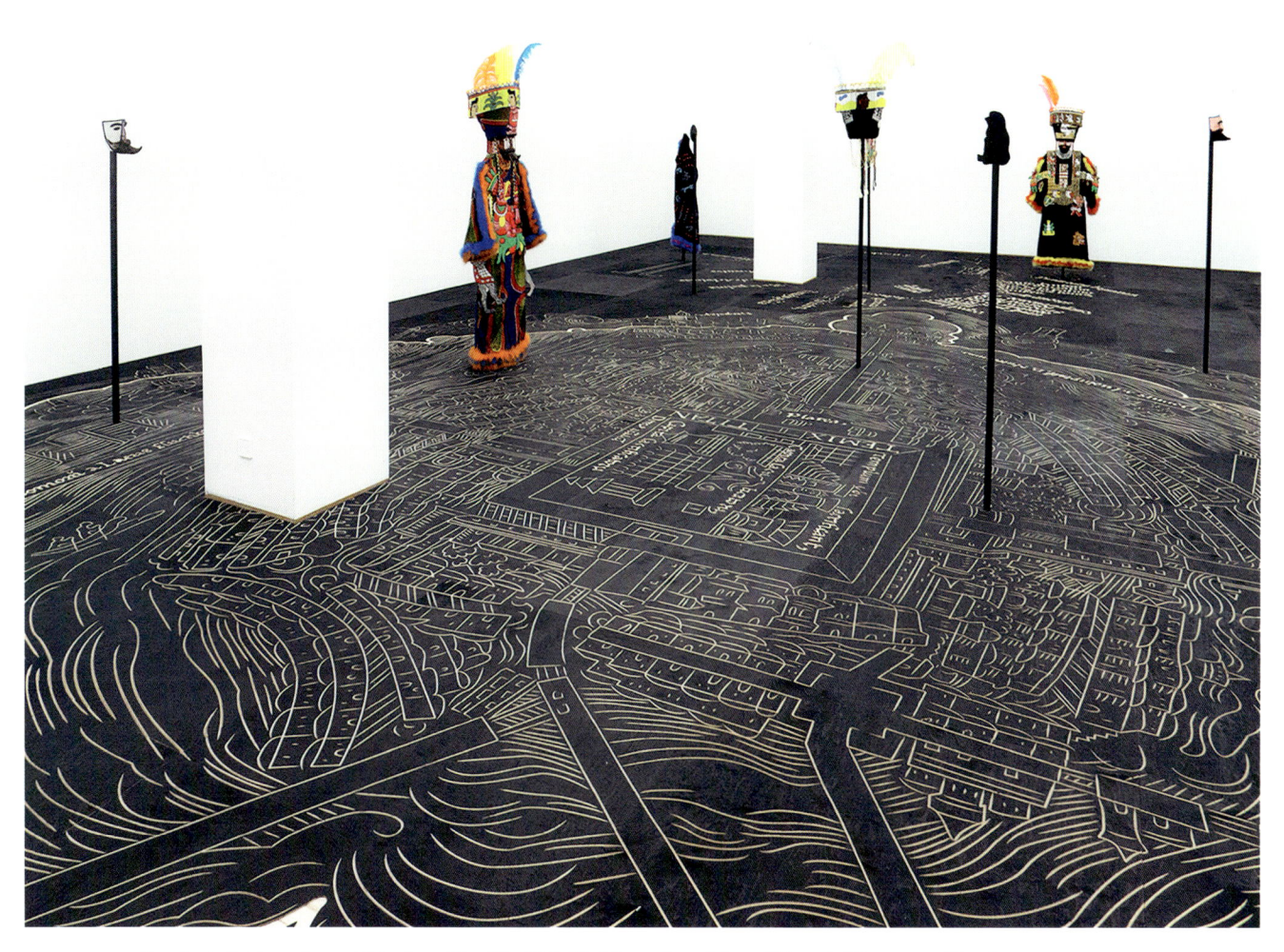

FIGURE 8.9. Installation view of Mariana Castillo Deball, *Nuremberg Map of Tenochtitlan*, 2013. Hamburger Bahnhof–Museum für Gegenwart, Berlin, 2013. Courtesy of the artist.

Because Castillo Deball's prints are so large, they render the map a frustrating object, because the partial prints withhold total knowledge. At the same time, for those who are deeply familiar with the map, partial knowledge can stand in for the whole, the prints serving as synecdoche. She calls the bound book an atlas, although unlike Braun and Hogenberg's *Civitates*, its individual pages were not composed so as to be intelligible. For me, though, the synecdoche raised another question: What exactly is the "whole" standing behind the fragment? It is the kind of question about systems of knowledge

and their representations that Castillo Deball's work frequently asks.

This work also complicates the typical modes of figuration. The dance costumes that are set on poles piercing the floor map are not those of Tenochtitlan's original inhabitants, nor of an imagined "Indian" of the kind painted onto the extra-illustrated map (see figure 8.6). Instead, these are costumes worn by participants in the *chinelo* dance—a spectacle first performed in the colonial era. During Carnival, a time of role reversals, residents of the modern state of Morelos would don masks and costumes and perform

FIGURE 8.10. View of Mariana Castillo Deball's Berlin studio, with Anna Szaflarski (*center*) and Ayami Awazuhara (*upper left*), 2013. Courtesy of the artist.

elaborate burlesques of their Spanish overlords. The *chinelo* dances continue today. In other words, the costumes are contemporary ones, worn by members of communities who might identify themselves as *pueblos originarios*. For Castillo Deball, the costumes may be meant to contrast the plurality of contemporary "Mexico" alongside the way that generic "types" appearing on maps reflect the ethnographic imperatives of an early modern European epistemology.

Another complication to figuration is produced by scale. Castillo invites visitors into the space of the map. But this move makes a second set of human bodies part of the artwork, at least for a time: the viewers who navigate around the Indigenous "presences" signaled by the costumes. Since the space of the map is not separate from that of the viewer, there is no comfortable place to view the map that does

not involve the movement of the body. This is far from the experience of armchair travelers looking at historic maps. For me, it was a disorienting experience, additionally so because walking on the piece involves some desecration: the dirt on the bottom of one's shoes leaves a mark with every step.

The meanings that these artists generate depend on the anteriority of the maps that they deploy. Both Rodriguez and Castillo Deball use well-known maps, which carry a host of built-in associations. Rodriguez's erasure of the California–Mexico border resonates precisely because one knows the importance of that border in defining where one nation's sovereignty ends and another begins. At the same time, it also captures the imagined territories of Chicanos, whose deep origins are in the Mexica homeland of Aztlán and who live in the United States, with cultural affiliations to both sides of the border. Concurrently, in asserting the natural world over the political order, Rodriguez also points to nature's ultimate sovereignty—a particularly relevant assertion in a time of climate change. And Castillo Deball taps into the familiarity of the 1524 map to produce unsettling effects that redound to other familiar cartographic objects.

Art is not a political program; nor does it offer a history lesson. At best, it can offer a kind of estrangement from the normal order of things, allowing us to return to that order with new eyes that are able to see fractures and seams. In the case of the work of Rodriguez and Castillo Deball, they helped me think harder about the close intersection between expressions of sovereignty, the history of cartography, and the long process of Indigenous dispossession in the Americas.[34] They also denaturalized some of those associations—particularly between expected modes of figuration and the map. As we look ahead, our familiar national map seems an outmoded instrument in confronting global challenges; weather pays no heed to national frontiers, and neither does the resulting flow of humans seeking survival. At the same time, maps are powerful tools because of their legibility and deep-seated authoritative claims. Perhaps we can learn something from artists about how to use them in new ways, to imagine pluralistic presents and to visualize futures that are more equitable and sustainable.

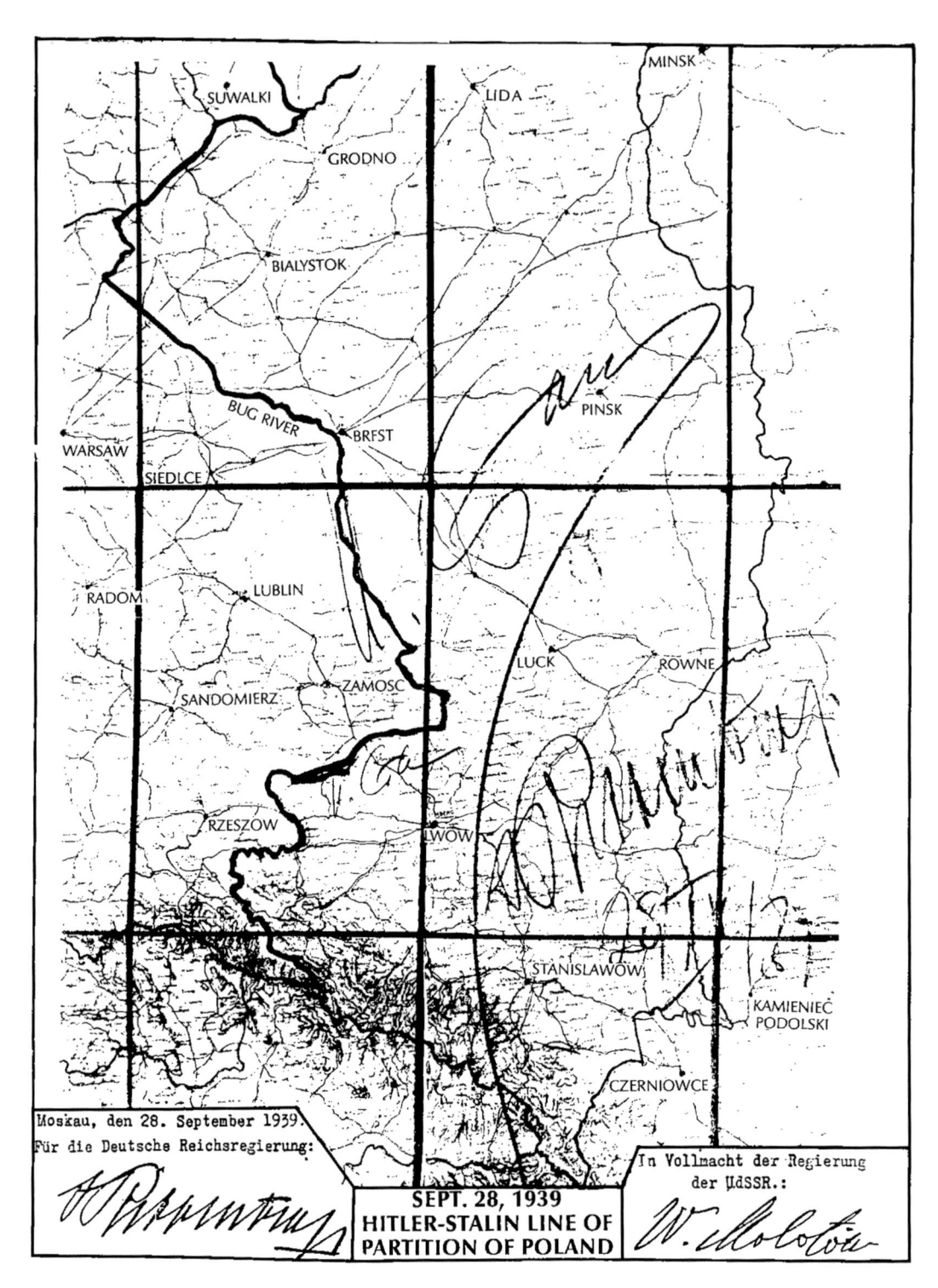

FIGURE 9.1. *Hitler–Stalin Line of Partition of Poland*, 1939. This map, an addendum to the 1939 nonaggression pact between Nazi Germany and the Soviet Union, demarcates the future spheres of influence of the two powers. The signatures of the foreign ministers, Joachim von Ribbentrop and Vyacheslav Molotov, are placed next to the boundary to affirm the agreement. Public domain. Available at https://commons.wikimedia.org/wiki/File:Molotov-Ribbentrop_Pact_showing_the_new_German-Soviet_border_Sept_28_1939.png.

9

Guntram H. Herb

Erasing the Other

Maps, Bordering, and Political Power

Mapping borders is key to exerting political power. The process of bordering—that is, defining the extent of the territory over which power is claimed, determining who and what is included and who and what is excluded—is difficult to conceive and convey without visualizations. When European explorers chanced upon the Americas, they had no knowledge of what lay beyond the coast, so in their greed to take possession of the new lands and their riches, they drew the imagined contours of the new continents on paper and dissected them with lines demarcating ownership. With the stroke of a pen, the fate of people and their lands were decided; in its most extreme form, even the world itself was divided between Portugal and Spain with a pope-sanctioned line of longitude in the Treaty of Tordesillas in 1494.

During the Berlin Conference of 1884/85, European leaders gathered in front of a giant map of Africa and divided the interior of the continent among themselves, scarcely knowing the places and peoples over which they claimed power. Though Europeans had explored some of the interior and shared tales of their journeys, such as the famed encounter between the journalist Henry Morton Stanley and Dr. David Livingstone, the spaces their maps defined were mostly blank.

The power of dissection by cartography is also vividly illustrated by the signatures of Joachim von Ribbentrop and Vyacheslav Molotov placed next to the boundary demarcating the future spheres of influence of Nazi Germany and the Soviet Union in the addendum to their 1939 non-aggression pact, a line that once more put a death knell to the Polish state (figure 9.1).

While many of these boundary-marking acts were carried out in closed meetings or agreed on in secret, as in the case of the Nazi–Soviet pact—power politics, plain and simple—they became a public affair with the advent of World War I, when the idea of national self-determination was elevated to a guiding principle for the peace that was to follow. It became

Research for this project was supported by Middlebury College. The Herder Institute for Historical Research on East Central Europe in Marburg, Germany was an essential resource; their map collection and staff are exceptional.

the cornerstone of the professed war effort of the Allied powers, France, Great Britain, and Russia. It enabled them to elevate the fight over Central Europe to a "just" war that was to give to the oppressed nationalities what was rightfully theirs.[1]

After the United States entered the war and President Woodrow Wilson outlined the conditions for the armistice and the Paris Peace Treaties in his famous Fourteen Points, the right to national self-determination was officially recognized for the first time by all major world powers.[2] With this new approach to bordering—redrawing the map of Europe along lines of nationality—also came a need for justifications and a wide public interest in maps of nations. Spurred by the new boundaries in the Paris Peace Treaties, such mapping became a public, mediatized, national affair, with each nation clamoring to use the evidence of "scientific" maps to back up their rightful claims to territory.

The public interest in maps intensified after the new boundaries reduced the territory of the largest country in Central Europe, Germany, by 13% and its population by 10%.[3] Soon, a veritable flood of maps of national territory appeared in Weimar Germany. Some featured novel forms of data, others innovative, persuasive design. At the same time German scientists leveled charges of fraud against French and Polish maps. While the rich trove of maps from this campaign offers deep insights into mappings of national territory, they have generally been dismissed in the literature as "unscientific" and mere propaganda because the German national boundaries presented in them helped justify Nazi Germany's conquests. But

where is the line between scientific and propaganda maps when recent scholarship considers all maps to be rhetorical?[4]

I will argue in this chapter that an analysis of the map campaign in the Weimar Republic holds promising insights for our understanding of maps of national territory. I carry out my investigation as follows: First, I look at the power of maps, their relationship to persuasion, and their emergence as evidence for national territorial claims. Next, I turn to the German map campaign: tracing its origins, critiquing its different initiatives, comparing its map designs and messages, analyzing its knowledge and persuasion networks, and assessing its impact. I conclude by discussing general lessons from the German case in the context of rethinking maps of national territories.

The Power of Maps and Maps of Sovereign Power

Maps are particularly powerful visualizations. Since the age of the pharaohs, maps have been associated with political authority and the administration of territory. Some argue that maps even predate language because they are an ideal way to present spatial information. In contrast to texts, which present information about places and their relationship to one another sequentially, maps portray places all at once, situationally. This allows maps to speak directly, even to those who are nonliterate. In contrast to paintings or photographs, maps are images that offer scaled and symbolic representations of the world.

We interact with them in a manner similar to the way we view landscapes, making them seem both familiar and trustworthy. The symbols they employ bring order to the complexity of the world and structure our understanding. As Anne Godlewska explains, "The precision of the lines on the map, the consistency with which symbols are used, the grid and/or projection system, the apparent certainty with which place names are written and placed, and the legend and scale all give the map an aura of scientific accuracy."[5]

The association of maps with truth and objectivity is further cemented by the epistemology of their practitioners. Cartographers stress the progressive nature of their enterprise to continually improve on accuracy, carefully differentiating their objective work from depictions that include distortions or lies. This binary distinction between objective scientific maps and deceptive propaganda maps is based on the communications model in cartography, which was dominant until the 1980s and still has a significant number of adherents today.[6]

Following this approach, the map is conceived as a true image of reality. The cartographer or sender of information seeks to represent the world in a way that allows the map user or receiver to comprehend the spatial information most effectively. Accordingly, cartographic researchers conduct surveys and experiments to determine how people read maps and which designs are most effective. Jacques Bertin's cartographic semiotics developed this further and created a grammar of signs and variables that took human visual abilities into account.[7] The stress on rules to produce efficient and effective communication is expressed in the "disciplinary apparatus" of scientific cartography: cartographic laboratories, technicians, projections, and design.[8] Propaganda maps are thus exposed as attempts to misuse this process to deviously create designs that convey a false image of reality to the map user. Examples include using colors to create positive or negative associations, selecting particular projections to emphasize particular places, focusing on particular areas, or omitting context.

In the late 1980s, these concepts were challenged in a fundamental way by the work of J. Brian Harley, who used ideas from Foucault and Derrida and considered maps as texts.[9] Unlike Bertin, he was not concerned with what could be "seen" in maps, but what could be "read" from them. Harley argued that maps reflect and reproduce the cultural and social context in which they were made. He deconstructed maps and pointed out, for example, that topographic maps show castles but not poorhouses, golf courses but not toxic waste sites. Only what seems important to society is mapped. Once the ideological loading of all maps was demonstrated, it was no longer so easy to distinguish between objective scientific maps that showed the "truth" and propaganda maps.[10]

More recently, Harley has also been criticized. Despite his critical attitude toward the orthodox positivist approach, he still believed that maps communicated unambiguous, clearly determined information about reality. While for Harley this reality was not "objective," it was still a social and cultural reality. For Denis Wood, Jeremy Crampton, and others, this does not go far enough. These geographers take a poststructuralist approach to argue that maps are more than pieces of paper used to communicate information about various places and spatial phenomena. Instead, they see maps as social constructions. As such—like all representations—maps can be interpreted in different, mutable, and contradictory ways.[11]

And yet maps appear to us as naturalized objects with a fixed meaning. How can this be explained? Judith Butler's theoretical work on performativity and reiteration, and further developments of her concepts by Rob Kitchin and Martin Dodge, are helpful here.[12] The continuous process of repeated use of maps and the common social context in which this takes place cement a particular meaning for particular maps. By context I mean the values and ideas that are dominant in our social milieu. We have limitations in terms of the information we can "see" in maps (which Bertin addressed). Here it is about the limitations and rules that define what we "read" from maps—that is, what we uncritically accept as given.

In this account, mapping should be understood as a process that is performative and reiterative—and maps should not be dismissed as biased solely on account of employing certain design elements. Such a perspective does not deny that some maps may be propaganda; on the contrary, it draws our attention to the fact that *all* maps offer only a particular and limited perspective on the world, as expressed in the saying "All maps lie flat, all flat maps lie."

The ability of maps to shape our perceptions of the world becomes uniquely relevant in claims to national territory and nationalism. As Benedict Anderson explained most succinctly, nations are imagined communities founded on stories of belonging to a group of people and to a particular place. A key question, then, is: How can the borders of this national land be defined in order to claim power over it? Landmarks like mountains, coasts, straits, and rivers convey a general idea but also leave much unspecified and unnamed. For example, the first verse of the German national anthem uses a visual language, referencing rivers and straits in all four cardinal directions; however, these do not meet up and thus leave considerable areas undefined. Only maps are able to communicate a precise image of the nation and foster a territorial consensus.[13]

In nations that emerged within existing state structures, such as France, territorial limits were pre-given and might appear less of an issue, but even there maps helped tie the population to the land. Inspired by Enlightenment ideals of individual freedom, secularization, and opposition to dynastic absolutism, it was the hexagonal shape of France that conveyed the new national space to the population, becoming a "logo map" in the aftermath of the revolution of 1789, to use Thongchai Winichakul's term.[14] When the idea of the nation spread to central and eastern Europe, where it followed ideas of Romanticism, it took on a different form. Romantic thinkers such as Herder and Fichte stressed the subjective, the emotional, the passionate in human beings and the need for the individual to find solace in the "organic" community that was made up of those who spoke the same language, shared the same folklore and customs, and were attached to the same native soil. Given the complex mosaic of ethnic groups in central and eastern Europe, with their ethnically mixed border zones and widely dispersed small settlements inside the main settlement areas of other groups, defining national territories was at once challenging and volatile.[15]

Thus, the first ethnographic maps across Central Europe appeared only in the 1840s, when national movements started to emerge. Diplomats took notice of the potential of these visualizations to incite demands for national liberation.[16] Initially, these new maps, such as *Sprachkarte von Deutschland* of 1844 by Karl Bernhardi, *Deutschland, Niederlande, Belgien und Schweiz: National-, Sprach-, Dialect- Verschiedenheit* of 1848 by Heinrich Berghaus, and *Nationalitäts-Karte von Deutschland* of 1848 by Heinrich Kiepert, were based only on historical records and regional descriptions of language use in different areas. Still, they spurred the national imagination. For example, Berghaus's designation of "Deutschland" (Germany) as an entity helped legitimize its distinct identity.[17] This new development did not escape the attention of the Austrian monarchy. Having had to use military force to reestablish control after the national uprisings of 1848, the Austrian government issued a large four-sheet map of the distribution of nineteen different nationalities in its territory. Authored by Carl Freiherr von Czoernig, the director of the office of administrative statistics, this *Ethnographische Karte der Oesterreichischen Monarchie* was the first such map based on census data. Morgane Labbé argues that its medley of colors suggests that identity based on language alone is not a solution, but rather that a multinational empire offers the possibility of unity in diversity.[18] More important, Czoernig's map initiated a new era of ethnographic maps based on official surveys. In the age of positivism, the use of scientific data gave additional credence to the information portrayed in maps. As products of science, they appeared objective and truthful—and could be presented as irrefutable evidence of the rightfulness of territorial claims. It was these types of maps that became central in the determination of new national boundaries at the Paris Peace Conference of 1919.[19] They also featured prominently in official revisions of borders such as the 1938 Munich Conference. The members of the German delegation brought such detailed, large-size maps that the carton holding them was too long to fit in their car and they had to drive with the top down.[20]

Mapping the German Nation

While the maps of Bernhardi, Berghaus, and Kiepert had already captured the German national imagination in 1848, concern with ethnic German borders among scholars making maps subsided after the German Empire was founded under Otto von Bismarck in 1871, becoming the domain of pan-German nationalists—chiefly geographers—such as Paul Langhans. For German geographers in the late nineteenth century, there was little to be gained in addressing the presence of ethnic Germans outside

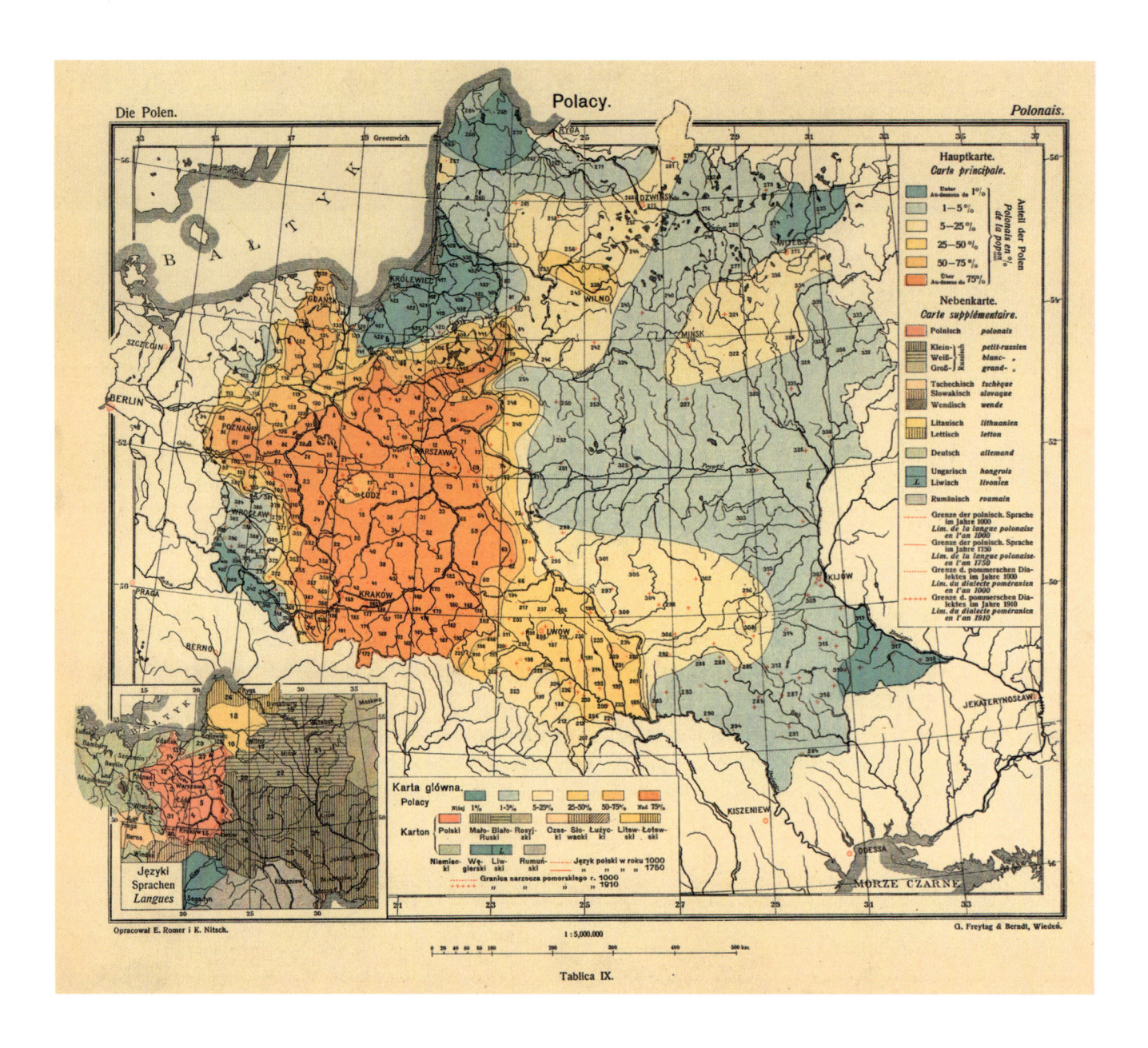

the boundaries of the Reich. These geographers were ardent supporters of the new German state, eager to receive recognition and financial support for their recently established university discipline.[21] However, confronted with the drastic losses in the Versailles Treaty half a century later, geographers quickly shifted their attention to German ethnic territory and searched for evidence to bring about a revision.

Their first activities were centered on finding explanations for why what was (in their eyes) unequivocally German territory would be severed from the German state. In this they laid blame on foreign maps—especially Jakob Spett's map of the distribution of nationalities in the eastern provinces of the German Empire and the ethnographic map in the 1916 atlas by the Polish geographer Eugeniusz Romer (figure 9.2).[22] Romer had studied with the renowned German geographer Albrecht Penck, who initiated scathing critiques of these maps.[23]

Penck pointed to examples of altered data, misleading color schemes, and unsuitable symbology in both Spett's and Romer's maps. He accused Romer of mixing the 1910 Prussian census with other survey data and wrongly subsuming minorities such as Kaschubes and Masures under Poles to arrive at a depiction that was most favorable to Polish demands.[24] The Spett map he labeled a "masterpiece of forgery."[25] Other geographers followed suit, calling the data representation in the Spett map a "scandalous falsification."[26] German geographers deemed the use of color in both to be biased, but they criticized Romer's ethnographic map in particular for showing data classes for the share of Poles above 5% in orange and those below 5% in blue. This gave the impression that all areas with more than 5% Poles were effectively Polish, which significantly increased the size of Polish claims.[27] A further point of contention was Romer's use of isolines (lines connecting points of equal values) to portray the distribution of Poles as percentages of the total population. Combined with the color scheme, this created a homogeneous Polish national core area whose boundaries faded out in smooth lines and extended uninterrupted along a corridor to the Baltic Sea. Penck argued that whereas the isoline symbology was useful to depict natural phenomena, such as elevation levels in topographic maps, it was misleading for languages or peoples because it obfuscated their intermixed distribution.[28] Perhaps most distressing to German geographers

was the realization that even existing German maps, such as those in the atlas of Langhans,[29] could be used against German claims since they showed sizable areas that were Slavic in the ceded northeastern territories.[30]

Recent scholarship on Romer's atlas by Morgane Labbé and Steven Seegel corroborates elements of the German critique, arguing that Romer combined different data sources and designs to create an image that was clearly intended to further Polish nationalist aims.[31] They also point to his disregard of minorities within lands he claimed for Poland, as well as his refusal to recognize Ukrainians as a nation, as evidence of national bias in his work.[32] This does not mean, however, that German geographers can claim the mantle of objectivity. They, too, reinterpreted official statistics in their critiques and claimed minorities such as the Kaschubes and the Masures to support their belief in the German character of the lost territories. Moreover, German maps employed the same kinds of misleading color schemes and data classes as Romer's. Dietrich Schäfer's 1910 language map of the eastern provinces represented all districts with 10% or more Germans in red,[33] and the same design trick was used to strengthen German demands to the Sudetenland at the 1938 Munich conference. In Erwin Winkler's map of German settlement areas in Czechoslovakia, even places with 20% German speakers could be claimed as rightful German territory.[34]

In one respect, the German critique provided important insights by drawing attention to a fundamental problem with maps of national territories.

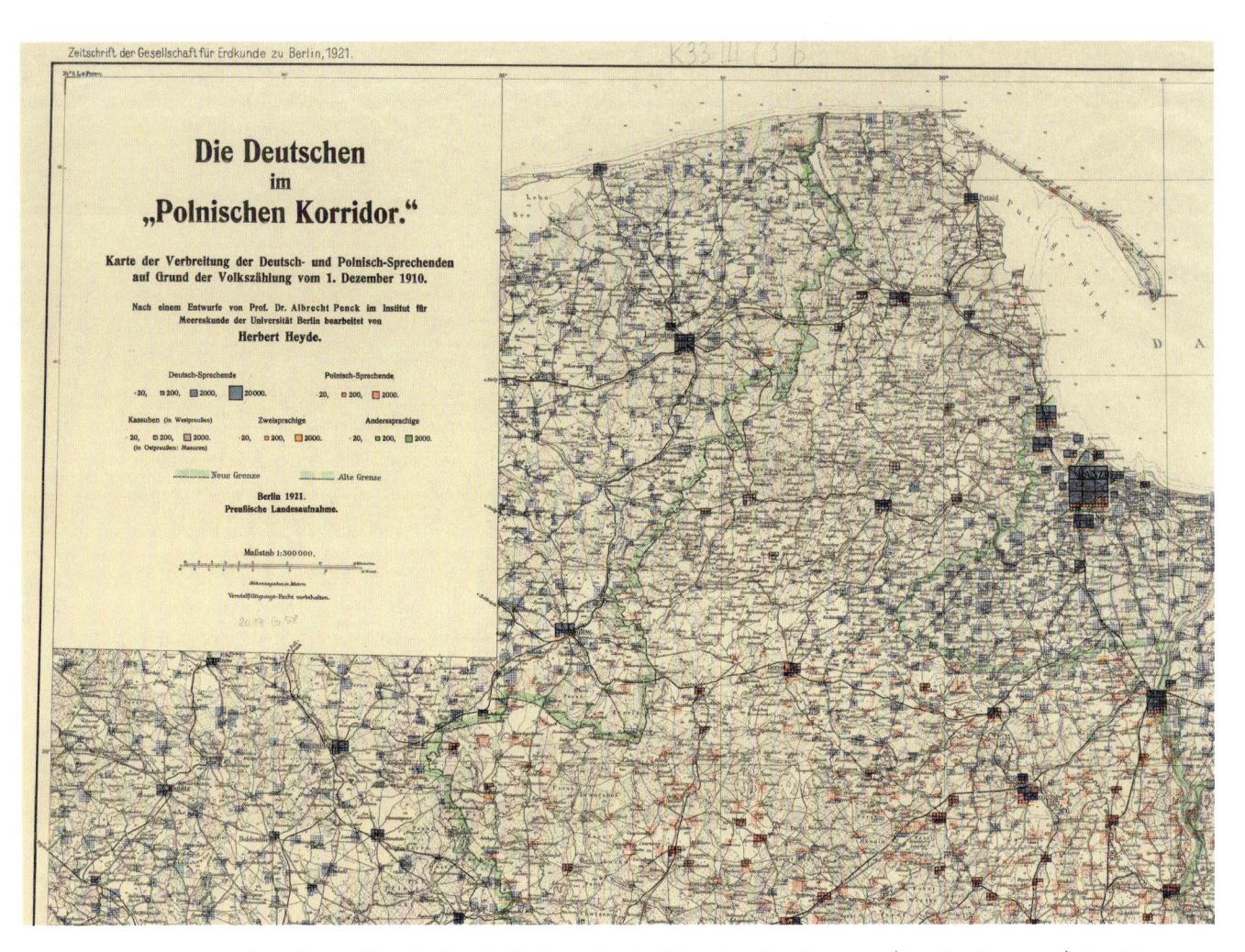

FIGURE 9.3. Detail from Albrecht Penck, *Die Deutschen im Polnischen Korridor*, 1921. (See also figure 9.6.) Penck's new design for an ethnographic map covered the ceded territory along the Vistula River at a scale of 1:300,000 and represented Germans, Poles, Kaschubes, Masures, bilinguals, and speakers of other languages (*Anderssprachige*). It depicted the absolute number of individuals for each group at their respective geographical location, with colored dots signifying 20 individuals, and colored squares of increasing size signifying 200, 2,000, and 20,000 individuals. Courtesy of Herder-Institut, Marburg, Kartensammlung: K 33 III C 3 b.

With the exception of Romer's isoline map—an unusual and rare case[35]—essentially all of them are choropleths: maps that distinguish different (in this case, national) territories by shading administrative districts in contrasting colors or hues. This results in clearly bounded territorial units, which aids the demarcation of national borders at the cost of erasing minority populations within each unit and disregarding population densities. In these kinds of maps, national minorities are subsumed under the majority color or hue, and urban and rural districts of similar area are given the same visual weight even if they contain populations of vastly different size. Albrecht Penck was again one of the first to point out this issue and to develop a new design for an ethnographic map. He first issued a small-size map in a newspaper[36] before presenting a full version to the scientific world in a scholarly journal.[37] Covering the ceded territory along the Vistula River at a scale of 1:300,000, Penck's map represents Germans, Poles, Kaschubes, Masures, bilinguals, and "speakers of other languages" (*Anderssprachige*). It depicts the absolute number of individuals for each group at their respective geographical location, with colored dots signifying 20 individuals, and colored squares of increasing size signifying 200, 2,000, and 20,000 individuals. As figure 9.3 shows, Penck's new design gave a true impression of the intermixed distribution and variation in density of nationalities. Yet for this very reason, as one critic pointed out, its portrayal of the dispersion of different groups made it exceedingly difficult to base boundary delimitations on them.[38]

Faced with the difficulty of using existing or newly developed ethnographic maps in support of a border revision, German geographers adopted a new approach: the development of new maps that used alternative definitions of national territory. The fact that Penck here once again took the initiative underlines his centrality in the ideological mapping of German national territory. His concept of the *Volks- und Kulturboden* (German national and cultural soil) received a prominent display: its map appeared in schoolbooks, journals, the press, and public lectures. In a transformative article and map of 1925, Penck explained that nations were deeply connected to and dependent on the land.[39] But not all nations were created equal. Some, such as the Germans, were most advanced and therefore had a powerful impact on the national landscape. Others, such as the Slavs, he deemed less capable of creating a lasting imprint. Thus, as a geographer, he could easily go on the ground and see where the carefully kept "advanced" German national land ended and the less improved Slavic landscape began. Wilhelm Volz summarized what he observed on landscape maps: "The culture is German, the forest is Polish."[40] Introducing such a national hierarchy allowed the map to claim all territories lost in the Treaty of Versailles and even lands beyond.

Geographers felt inspired. Other scholars suggested that Germans could claim even more, pointing to elements of the German cultural mission in the East: the use of German in commerce, excursions of historic personages, the spread of German law, and the like. Hermann Lautensach, a student of Penck, presented a garland of cultural border lines.[41] When Penck's original map was reissued with the help of Hans Fischer, both as a color wall map and as a page in school atlases, they added the category "Use of German in Commerce" (figure 9.4).[42] This allowed them to visually lay claim to an enormous area in the East—one that would correspond surprisingly closely with the territories conquered later by the Nazis.

As it turns out, German concern with map falsifications and the creation of new maps was not an affair of isolated individuals but a widespread social movement, driven by shared grievances about the new borders.[43] It was an assemblage of people, ideas, places, and objects that came together in an impactful way through two interlinked networks—one cen-

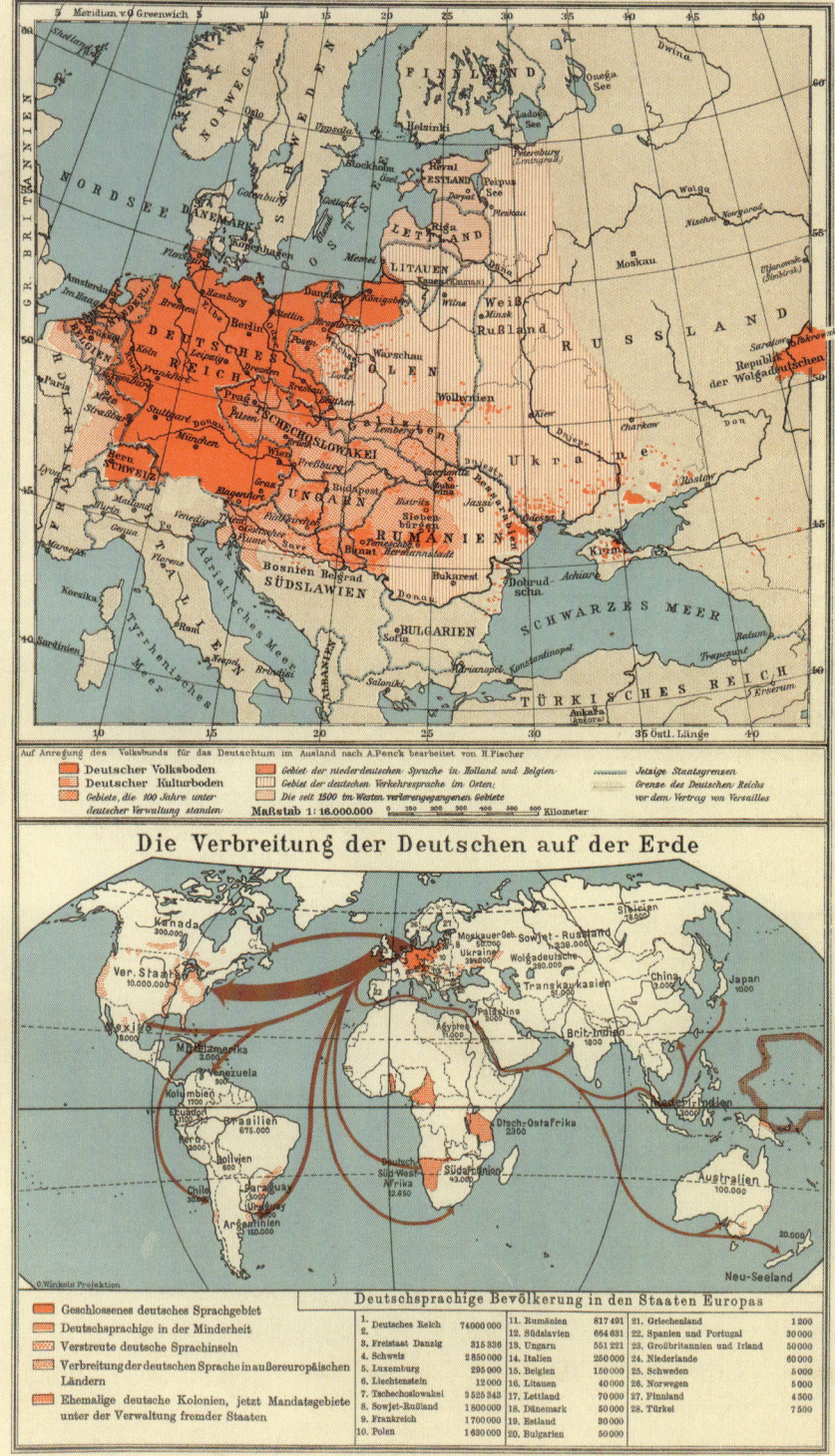

FIGURE 9.4. E. Debes, *Handatlas, Der deutsche Volks- u Kulturboden in Mittel- u Osteuropa*, from *Columbus Volksatlas* (Columbus Verlag, Paul Oestergaard K.G, 1938). This map from a school atlas presents territories imprinted with the "national and cultural soil" of Germans based on Albrecht Penck's concept of *Volks- und Kulturboden* and the added category "Use of German in Commerce." Courtesy of Herder-Institut, Marburg Kartensammlung:- 2° 20.03968, S. 55b.

tered on the production of knowledge, the other on effective communication.

From the beginning, geographers like Albrecht Penck and Wilhelm Volz did not just study visualizations of national territory, but collaborated with pan-German nationalists, such as Karl Christian von Loesch, to develop scientific evidence in support of a revision of the Treaty of Versailles. They strategized in secret conferences, socialized at the Deutscher Klub in Berlin, and secured funding for research on German influence and heritage in Eastern Europe. A central node of this knowledge network was the Mittelstelle für zwischeneuropäische Fragen in Leipzig, which coordinated conferences, initiated and funded research activities, and authored and edited data collections and publications in order to arrive at a common understanding of what was rightfully German territory.[44]

It was in the context of these network activities that Albrecht Penck developed his *Volks- und Kulturboden* (ethnic and cultural lands) concept. The Mittelstelle adopted this wording as part of its name after 1925, becoming the Stiftung für deutsche Volks- und Kulturbodenforschung. This organization acted as an enforcer of a unified stance on depictions of German national territory.[45] When publications that depicted ethnic distributions unfavorable to German interests came to the attention of the Stiftungs director, Wilhelm Volz, he informed government authorities or threatened the publishers with boycotts and legal suits. One publisher was forced to reprint a map for free to give to all its atlas customers because it had omitted German place names on a map of Czecho-

slovakia. In the end, most major publishers, such as the Bibliographisches Institut in Leipzig, made sure they consulted with the Stiftung before publishing their maps.[46]

Mapping of German national territory also was part of a persuasion network. Three strands came together in its emergence. First, German scientists who studied the accuracy of ethnographic maps had noticed that foreign maps paid more attention to effective presentation, which piqued their interest in more persuasive designs. Second, followers of German *Geopolitik*, such as Karl Haushofer, promoted the use of suggestive maps to convince the German people that a disregard of geographic conditions had led to their country's downfall and that only a sound knowledge of geopolitical realities would bring about a return to world power status.[47] Third, nationalists across the German right embraced the myth of "the stab in the back," or *Dolchstosslegende*. This was essentially the idea that the German people had been demoralized by superior enemy propaganda and led to sign a humiliating peace treaty even though their armies had been victorious on the battlefield.[48]

The lesson in every case was clear: if Germany wanted to regain lost territories and become a world power again, it needed more effective and convincing maps.

Though many individuals were involved, one cartographer emerged as a key figure: Arnold Hillen Ziegfeld. A graphic designer and book publisher by trade, Ziegfeld tirelessly promoted the use of maps as tools of persuasion. He published articles, submitted petitions to the government, and established map-

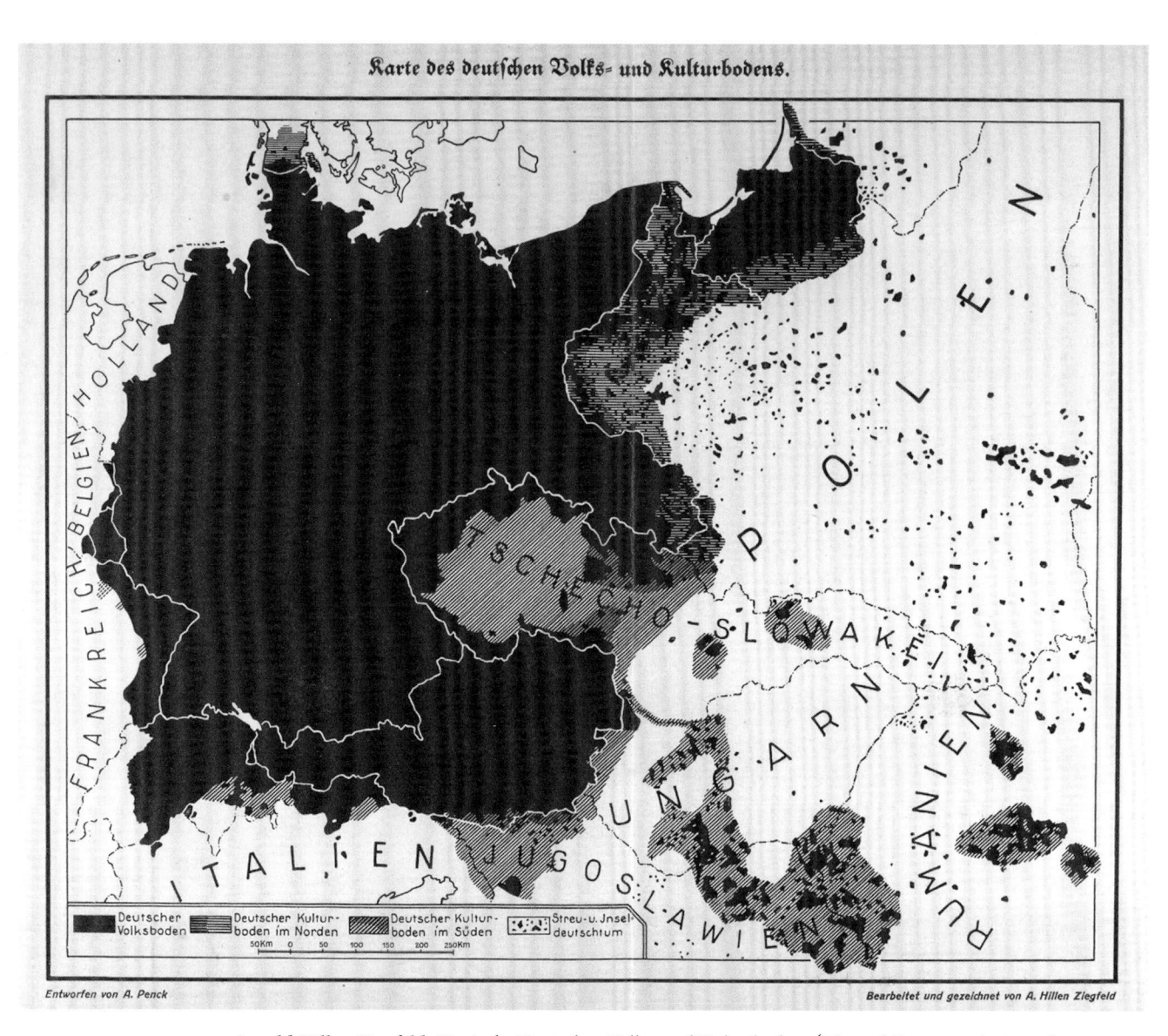

FIGURE 9.5. Arnold Hillen Ziegfeld, *Karte des Deutschen Volks- und Kulturbodens* (Map of German ethnic and cultural lands), 1925. This simple and captivating map contributed to the immense influence of Penck's *Volks- und Kulturboden* concept. Public domain. Available at https://digital.library.cornell.edu/catalog/ss:3293845.

design offices in various publishing houses. The new maps that he helped create employed designs that were simple, captivating, and compelling (see figure 9.5). It was Ziegfeld's masterful cartography that contributed to the immense influence of Penck's *Volks- und Kulturboden* concept.[49]

As constituents of these networks, maps of Ger-

man territory had agency in two ways. They functioned both as a Foucauldian technology that was directly implicated in the control of territory and as an opportunity multiplier that enhanced social mobilization around the idea of a Greater Germany. First, data-based ethnographic maps played a key role in the coordination and standardization of knowledge. As what Bruno Latour calls *immutable mobiles*,[50] they enabled the transfer of knowledge about German national territory to central institutions, where the information was collected, interpreted, mapped, printed, and then sent back across the network as exact copies of the collated version to academic institutes, government agencies, military offices, and businesses. The information contained in the map was a prescription for action; users were directed in their actions by the information contained in the map. For example, the military made decisions on where to attack or round up civilians based on how the ethnic terrain was depicted. The map became a surrogate for the territory, facilitating control at a distance. This explains why the SS gathered foreign ethnographic maps whenever they came across research institutes or libraries, bringing these captured documents back to the offices that were engaged in planning the reorganization of the occupied eastern territories.[51] The information on the distribution of different nationalities was then used in the strategy of divide-and-control that is the hallmark of empire.

Second, persuasive maps of national territory had a powerful visual appeal that drew readers in and delivered an effective message. Their simple design made them accessible to a wide range of people, which meant greater exposure and the chance to win increasing numbers of converts to the nationalist ideas that were presented. These maps were also ideally suited for education. As schools are one of the pillars of nation building, schoolroom maps are instrumental in conveying the border between us and the Other that is at the heart of national identity.[52] It is particularly effective to introduce national maps to children at the age when they are developing their spatial identity.[53] This has lasting effects as knowledge is instilled at the beginning of a new generation. The networks behind the production and dissemination of maps thus reinforced one another, using both data-driven and persuasive maps to establish a hegemonic national discourse of the "true" shape of Greater Germany. The resulting national consensus was instrumental in justifying Nazi expansionism.

Rethinking Maps of National Territory

The German map campaign shows that maps definitely are not "epistemically groundless," as Seegel claims.[54] On the contrary, their agency and power helped define and control the ground—even at a distance. Maps of nations, especially choropleth and isoline maps, are spatial frames that allow us to observe the world selectively and reduce its complexity. By their very design—areal shading to identify and distinguish areas of national majorities—they *silence* variation within, constructing an identity of the *Other*, and *conflate* scales to the national. They thus have the main constitutive elements of what critical geopolitics scholars call hegemonic "scripts." These maps act on us by conveying explicit arguments; they are human creations, even if their scientific idiom persuades us otherwise. Maps reinforce the misconception of distinct national territories and reify their existence. As Latour explains, maps make it possible not only to *see* the invisible, but also to prove its existence: an existence that can be controlled, since it is movable across time and space.[55]

German academic geographers had a powerful influence on conceptions of national territory through such immutable mobiles. They provided scientific legitimacy to the national cause, and the evidence they collected appeared objective and trustworthy. They also had a direct influence on the classroom, since academic geographers trained the nation's teachers. But mapping national territory was not just a German obsession. Geographers all across Europe, including Eugeniusz Romer, Emmanuel de Martonne, and Jovan Cvijić, were engaged in the same enterprise.[56] They all worked to advance their nation's interests, using networks of knowledge to help shape ideas into a powerful message. The same dynamic is visible today—and given the rise in ethnic nationalism across the world, there is little hope that visualizations of national territory will fade away. The continuing allure of territory, as Alexander Murphy has so aptly put it, makes maps indispensable.[57]

On a brighter note, these efficient tools of control are not irrevocably tied to hegemony; they can also be turned into tools of destabilization and critique. Penck's dot maps in particular are inspiring (figure 9.6).[58] Even though they were intended to help the German cause, they nevertheless can be seen as subversive, for by foregrounding the intermixed distribution of nationalities in borderlands, they highlight the difficulty of drawing borders. Their focus on

points foreshadows the pointillism of digital mapping, which is currently being used to deemphasize territorial limits (as discussed in this volume by Luca Scholz).[59] If seen in this way, Penck's maps can help destabilize our obsession with the sanctity of border lines and homogeneity.

Similarly, Eugeniusz Romer's nationalist isoline mapping style can be instrumentalized to efface the settler colonial logic of US and Canadian maps (figure 9.7). Official Canadian maps of "aboriginals" all end abruptly at the border, as if no Indigenous peoples lived outside their beloved country. US maps of "Indian Country" practice the same cutoff at the border and conveniently encase the original inhabitants in reservations, thereby silencing 70% of Native Americans not living there. The segregation extends to nomenclature and legality: First Nations and Native Americans; reserves and reservations; status cards and tribal ID cards. The two countries also differ in their definitions of indigeneity and their recognition of Indigenous crossing rights, such as those enshrined in the Jay Treaty of 1794. The counter-map in figure 9.7 allows us to envision an alternative to the world of settler states: one where a contiguous Indigenous presence and exchange across North America overwrites hegemonic linear borders. Together, figures 9.6 and 9.7 imagine a world that is not defined by the patchwork of exclusive, uniform national territories; they dream a remapped space of mixture, fluidity, and exchange.

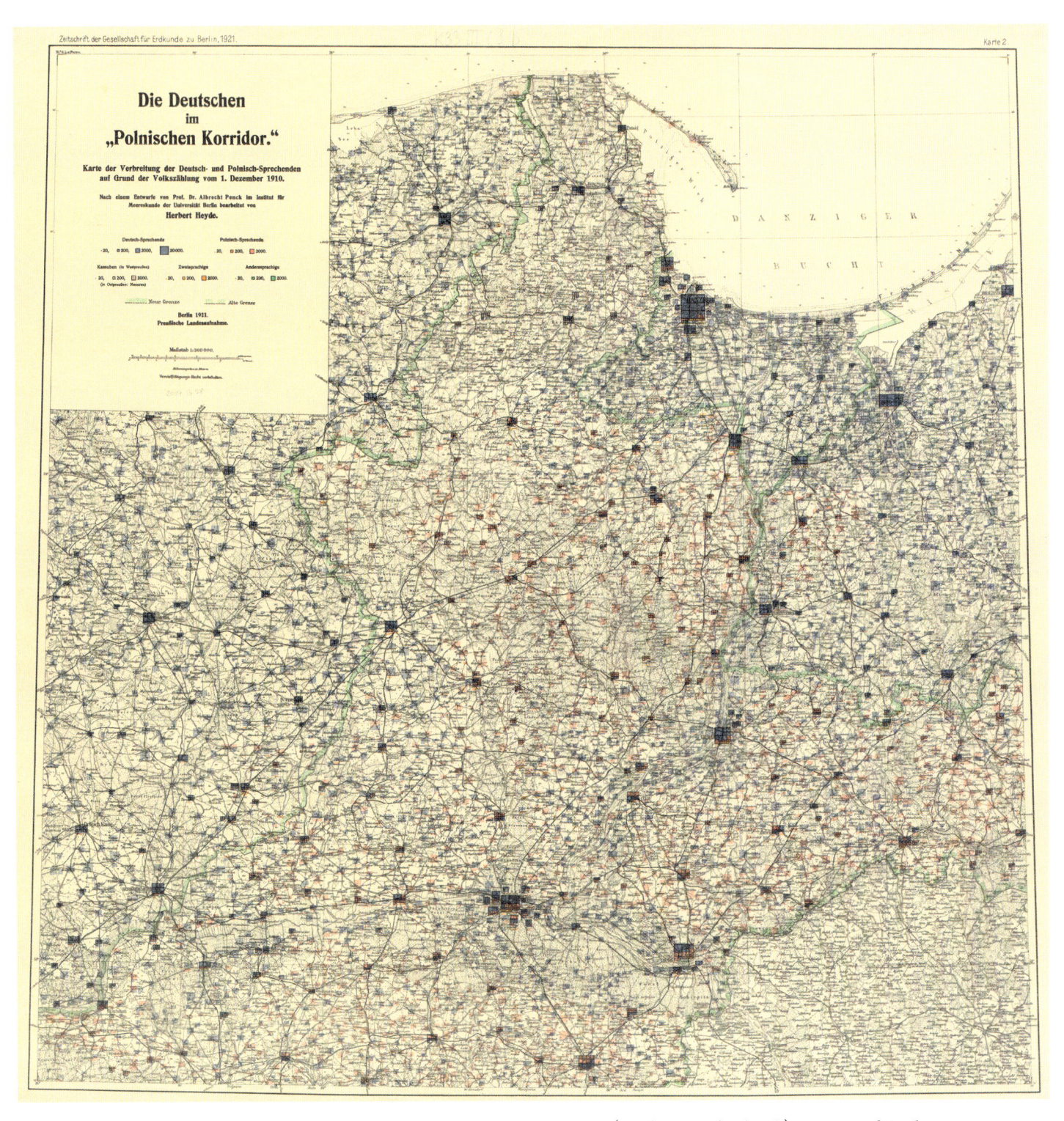

FIGURE 9.6. Albrecht Penck, *Die Deutschen im Polnischen Korridor*, 1921. (See figure 9.3 for detail.) Courtesy of Herder-Institut, Marburg, Kartensammlung: K 33 III C 3 b.

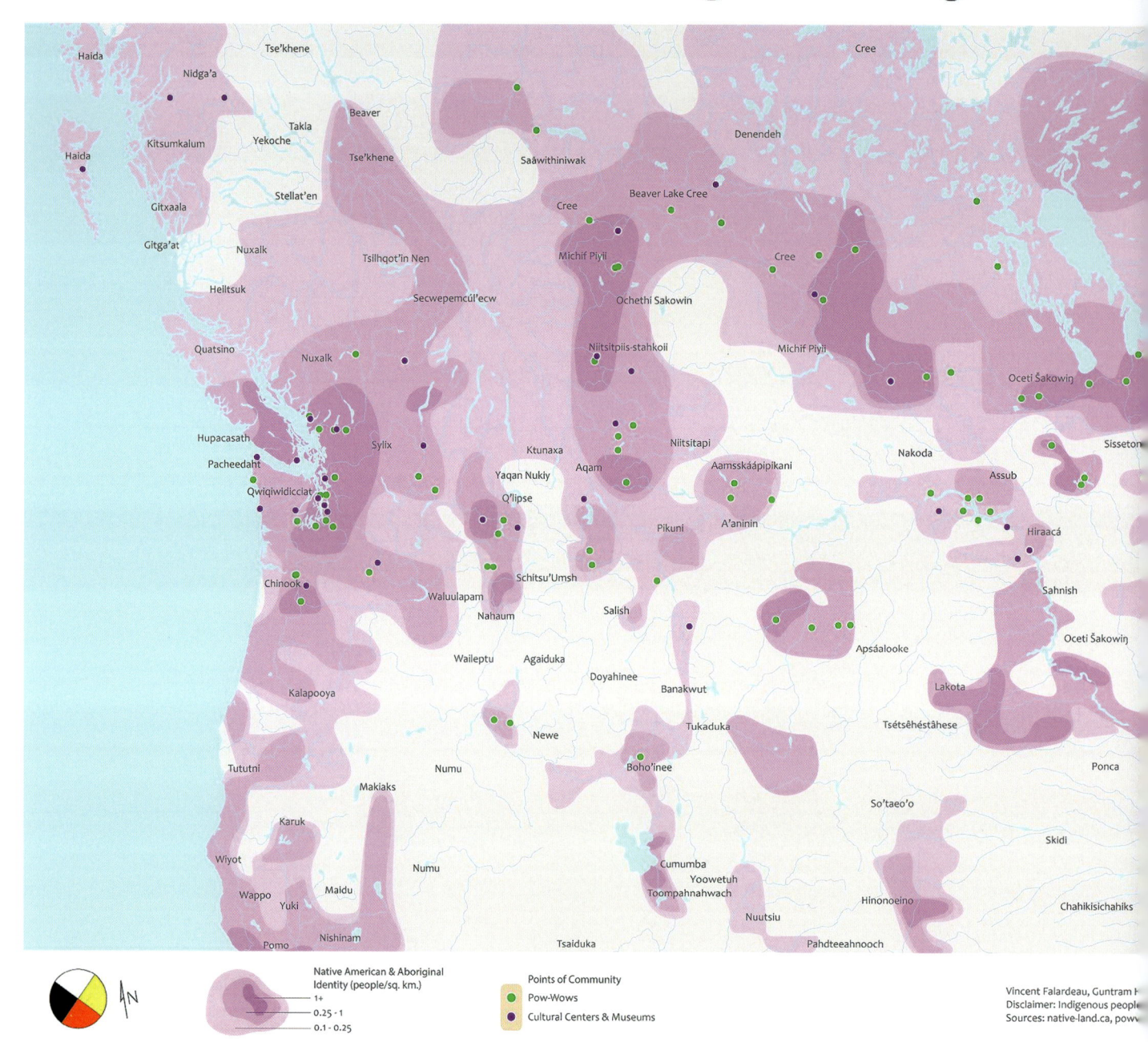

FIGURE 9.7. Vincent Falardeau, Guntram H. Herb, and Kathryn Talano, *Erasing the Line: Indigenous Community across the US–Canada border*, 2023. This map is part of a larger project on the Indigenous borderlands along the US–Canada divide (www.border-rights.org). It is a cartographic exploration from within the settler episteme that seeks to counter the almost universal silencing of a contiguous Indigenous presence and exchange across the border in official maps as well as in popular scientific publications (see, for example, *National Geographic*'s map of "Indian Country"). The design of the counter-map is inspired by Eugeniusz Romer's innovative use of isolines for the depiction of national groups. Rather than confining different concentrations of national groups within the irregular and

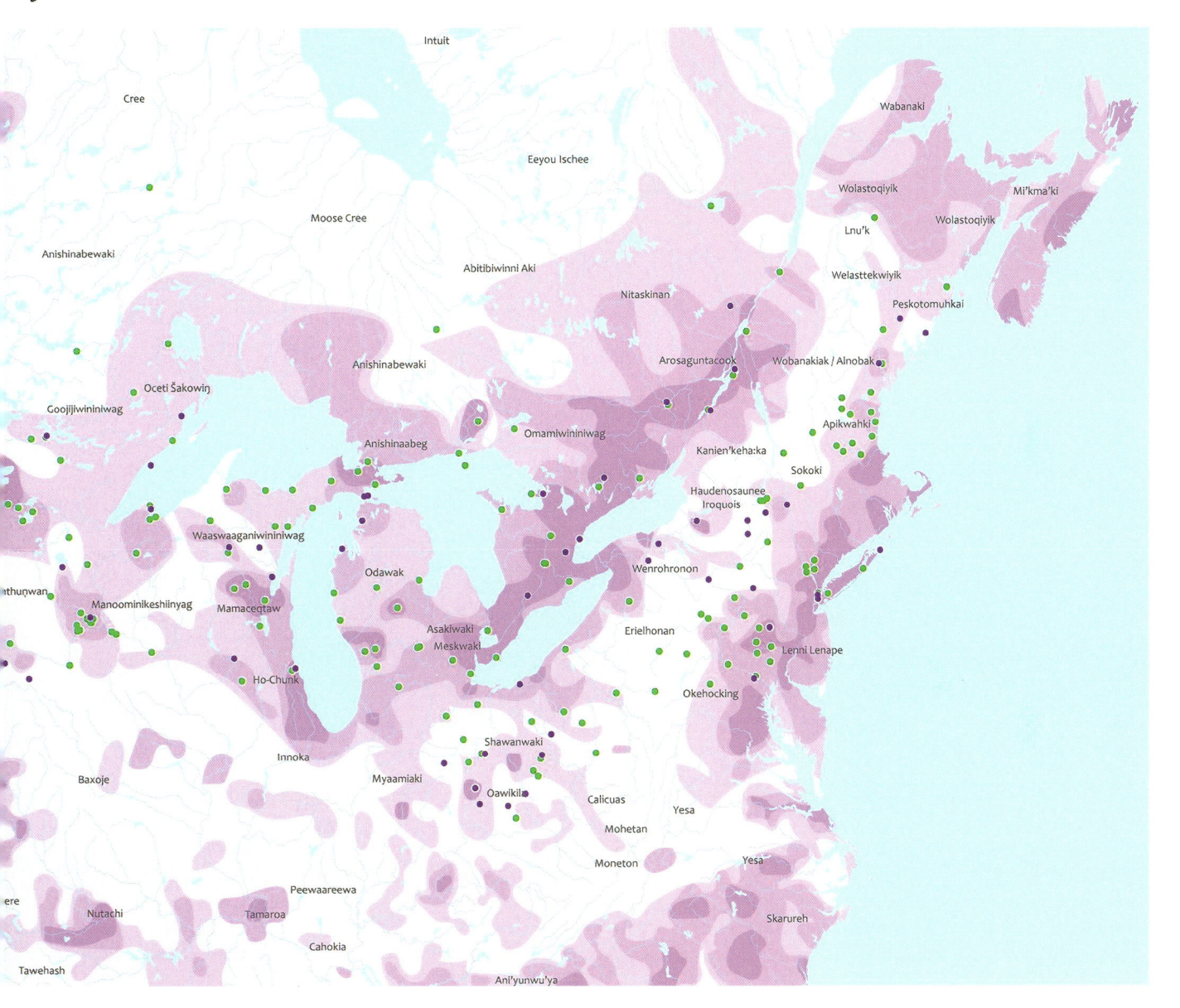

Intuit

Cree

Eeyou Ischee

Moose Cree

Anishinabewaki

Abitibiwinni Aki

Nitaskinan

Wabanaki

Wolastoqiyik

Lnu'k

Mi'kma'ki

Wolastoqiyik

Welasttekwiyik

Peskotomuhkai

Oceti Šakowiŋ

Goojijiwininiwag

Anishinabewaki

Anishinaabeg

Omamiwininiwag

Arosaguntacook

Wobanakiak / Alnobak

Apikwahki

Kanien'keha:ka

Sokoki

Haudenosaunee
Iroquois

Waaswaaganiwininiwag

Odawak

Wenrohronon

athuŋwan

Manoominikeshiinyag   Mamaceqtaw

Asakiwaki
Meskwaki

Erielhonan

Lenni Lenape

Ho-Chunk

Okehocking

Innoka

Shawanwaki

Baxoje

Myaamiaki

Oawikila

Calicuas

Yesa

Mohetan

Moneton

Yesa

Skarureh

Peewaareewa

ere

Nutachi

Tamaroa

Cahokia

Tawehash

Ani'yunwu'ya

...ed without intimate knowledge of their history. Spellings may reflect varied sources. Isolines of population have been drawn freely from county choropleth maps, with liberties taken for the purposes of simplicity and cartographic style.
...M Elevation Data, Natural Earth, US ACS 2019, Canada Census 2016

arbitrary boundaries of an administrative district, isolines show smooth gradations of concentrations that reflect the contiguous transborder Indigenous presence of cultural centers, powwows, and community names. This map does not intend to represent official or authentic Native community interests and should not be viewed as an example of Indigenous counter-cartography. The design of the map is explained in a research note in *Borders in Globalization Review*, which also includes a link to a full-size image of a slightly different version, available at https://www. davidrumsey.com/luna/servlet/detail/RUMSEY~8~1~342764~90110908:Erasing-the-Line--Indigenous-Commun.

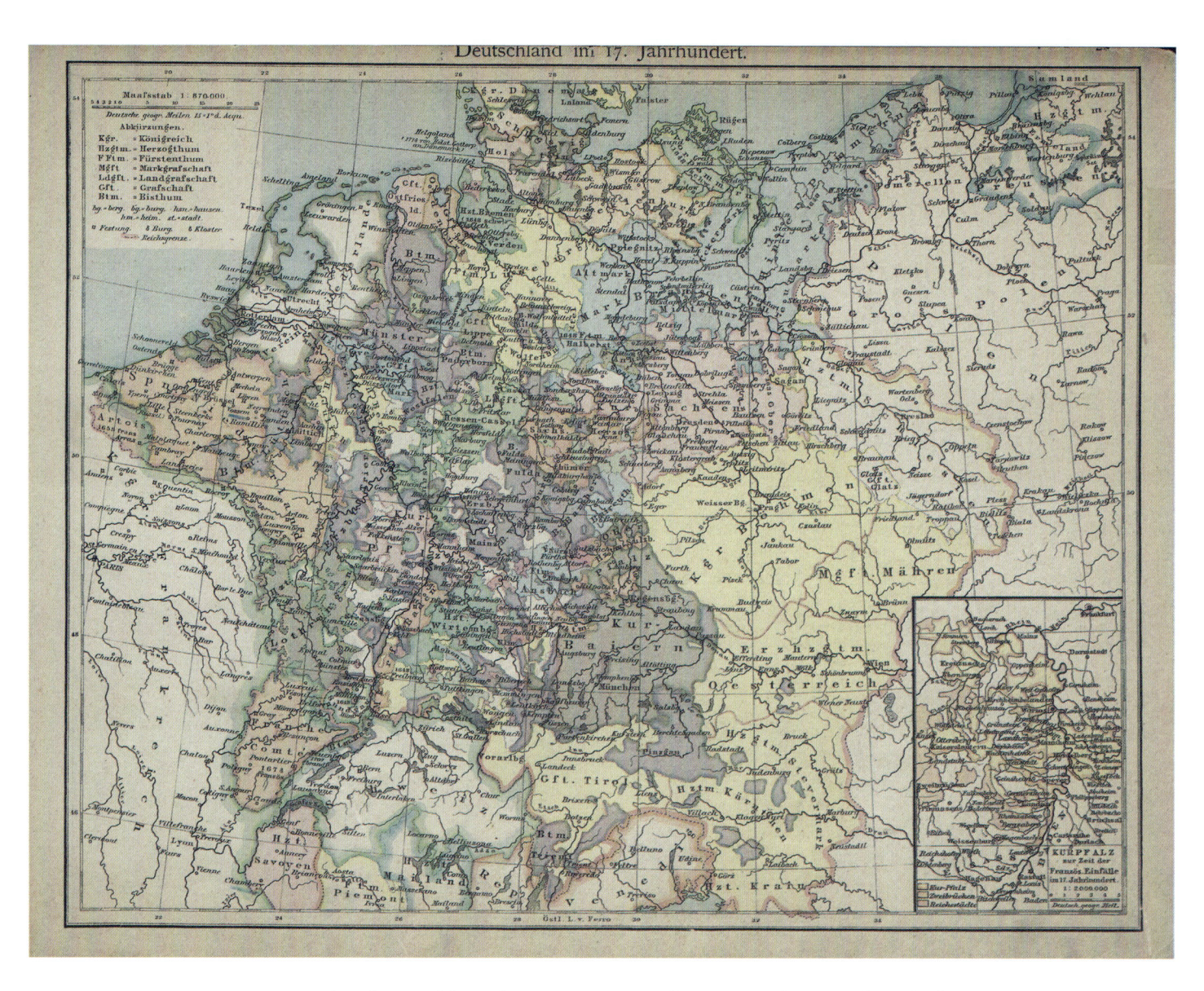

FIGURE 10.1. A patchwork map of the Holy Roman Empire in the seventeenth century from the first edition of Friedrich Putzger's historical atlas. The same atlas is at the basis of many maps of the empire found on Wikipedia today. "Friedrich Wilhelm Putzger, *Deutschland im 17. Jahrhundert*," in *Putzger's historischer Schul-Atlas zur alten, mittleren und neuen Geschichte* (Bielefeld: Velhagen & Klasing, 1877), 21. Public domain.

10

Luca Scholz

Visualizing Shared Dominion in the Holy Roman Empire

Dilution, Orientation, Oscillation

Scholarship on the premodern German lands can evince a sense of "cartographic lag." For more than half a century, historians of the Holy Roman Empire of the German Nation have revised many of the negative evaluations of the empire that were common in the historiography of the nineteenth and early twentieth centuries. Historians began to emphasize, for example, the empire's effectiveness at ensuring peace, religious coexistence, and the independence of its least powerful members.[1] The younger historiography has also deployed a range of terms to describe the empire's distinctive spatial-political makeup that point beyond the image of a fragmented political landscape, including old and new concepts such as associative political culture,[2] shared sovereignty,[3] *territorium inclausum*,[4] or fractality,[5] to name but a few. This varied and evolving conceptual repertoire stands in sharp contrast to the relatively stable graphical conventions with which the empire is commonly represented on maps. With important exceptions, thematic maps of the empire—such as figure 10.1—continue to emphasize contiguity, exclusion, and self-containment, while deploying a more muted graphical repertoire when it comes to visualizing the manifold forms of spatial competition, cooperation,

and indeterminacy that were characteristic of political life in the empire. The aim of this chapter is to examine how such nonexclusive forms of dominion have been and can be represented on retrospective historical maps.

Interest in alternative modes for visualizing the spatial-political complexity of early modern Central Europe ties in with broader methodological discussions in spatial history. Recent decades have witnessed an increased uptake of digital mapping as a method of historical research.[6] Spatial historians have invested considerable conceptual and technical sophistication into developing novel and creative ways of collecting, processing, and analyzing historical data.[7] This chapter aims to highlight the interest and importance of cartography and visualization for spatial-historical argumentation. In this, I follow others who have highlighted the importance of inductive, qualitative, and topological forms of visualization and spatial representation in the humanities.[8] The premise of this work is that maps, like any other visualization, are not simply surrogates for data but visual arguments which rely as much on graphical choices made in the process of visualization as on the underlying data itself. Many scholars reject the view that data speak for themselves and instead emphasize the interpretative choices at work when data is collected, organized, and displayed.[9] Scholars and practitioners of data visualization and computer graphics are proposing creative ways to counter positivist or totalizing impulses.[10] While spatial historians traditionally emphasize the experimental dimension of mapping as a form of layering and exploring spatial data, the cartographer and historian William Rankin recently argued that "instead of simply discovering patterns in data, the goal should be to find visual strategies for seeing the world differently."[11]

It is in this spirit that this chapter proposes cartographic strategies for representing forms of dominion that are not in wide currency in contemporary historical scholarship. In the early modern German lands, the spaces over which rulers exercised their rights and privileges—such as jurisdiction, taxation, safe conduct, or hunting—did not always coincide.[12] The coexistence of hundreds of fragmented, fuzzy, and often overlapping political entities that were integrated in a common imperial structure make the Holy Roman Empire an interesting laboratory for studying models of shared power. While some imperial estates at least partially succeeded in rounding up their possessions into a more homogeneous *territorium clausum*, variegated forms of territorial complexity persisted well into the nineteenth century. This chapter will focus in particular on forms of nonexclusive dominion—that is, instances where competing polities exercised authority over the same spaces.

Following a discussion of contiguity and fragmentation, which comprise prominent yet frequently questioned tropes of conceptual and visual representations of the Holy Roman Empire, this chapter presents three alternative strategies for mapping shared forms of dominion. The first section argues that the political density suggested by patchwork maps can be reframed as a dilution of authority. The second section introduces a map that emphasizes the linearity of political claims over road infrastructure—a context that engendered a range of shared, nonexclusive, or competing forms of dominion, including separation by orientation. The final section presents the case of a diachronic (alternating) condominium, in which high jurisdiction was both shared and exclusive. Adding to the design repertoire introduced in the previous chapter by Guntram Herb, these maps together offer concrete visual strategies for redrawing shared forms of dominion in and beyond the German lands.

Visual Metaphor: Beyond Contiguity and Fragmentation

Efforts at mapping the Holy Roman Empire's political divisions as a patchwork of contiguous polygons are sometimes questioned on practical grounds. Joachim Whaley, for example, has lamented the "technical impossibility of reproducing the internal territorial boundaries on the page in such a way that the eye can make sense of their sheer complexity."[13] The challenges involved in collecting, processing, and visualizing geospatial data on the political geography of the imperial estates are indeed daunting.[14] However, many scholars go further and argue that mapping the

Holy Roman Empire as a landscape of fragmented, bounded political entities is not just technically challenging but conceptually problematic, in that this practice effectively frames the empire's political geography in ways that are misleading. In the face of "overlapping lordships and an imponderable nexus of rights, claims, and aspirations," Whaley for one found the insistence on frontiers "anachronistic."[15] Falk Bretschneider and Christophe Duhamelle likewise argued that the notion of fragmentation tends to obscure the connections between polities and suggests an "aleatory" splintering without internal coherence.[16] In a similar vein, Wolfgang Wüst has suggested that patchwork maps distort perceptions of the empire as riven by chaos and conflict.[17] Without rejecting the notion of fragmentation, Peter Wilson warned against equating it with decline; in the history of the empire, "integration" and "demarcation" are not necessarily opposites.[18] Scholarly discontent with contiguity and fragmentation also ties in with critiques of territorial paradigms that emphasize "non-space-contingent ways"[19] in which power was exercised and conceived in the late medieval and early modern Holy Roman Empire.

That the visual metaphors of contiguity and fragmentation should be considered with caution is also suggested by studies in the history of cartography. Jordan Branch has argued that the cartographic preference for boundedness, homogeneity, and exclusivity initially bore little resemblance to the spatial organization of rule in early modern Europe.[20] Andreas Rutz emphasized the continuities between medieval and early modern forms of dominion and spatial inscription, with pointillist and areal forms of dominion coexisting.[21] While boundaries held an increasingly important place in manuscript and printed maps at different scales, these should not be taken as a straightforward index to changing political realities. On early modern print maps of the Holy Roman Empire, hand-colored boundaries became increasingly prominent, but map authors did not always attempt to represent political divisions exhaustively or in detail.[22] Early modern cartographers also regularly employed labels, coats of arms, crests, and ornaments rather than boundary lines to visualize political affiliation.[23] Boundaries played a more prominent role in manuscript and printed maps at the territorial or regional scale (especially in those commissioned by many imperial estates since the sixteenth century), and in the eighteenth century, maps established themselves as the key medium for the spatial description of boundaries.[24] Since the nineteenth century, popular historical atlases, such as Friedrich Wilhelm Putzger's (figure 10.1), opted to represent the Holy Roman Empire as a patchwork of fragmented principalities, but many authors continued to visualize the empire either as a coherent unit or as divided into dynasties, religious denominations, imperial circles, or other, larger entities.[25] It is also important to note that maps never fully replaced oral, material, and symbolic forms of inscription, as Andreas Rutz has shown. These other media were often less concerned with singularity and exclusion.[26]

The problem is, of course, neither exclusive to the Holy Roman Empire nor limited to the discipline of history. Other contributions in this volume provide striking examples, from imperial Russia—whose "indeterminate sovereignty" Valerie Kivelson describes with the metaphors of lacework, holes, and tenuousness—to the contemporary cases described by Franck Billé. John Agnew famously called the tendency of international relations theorists to conceive of state territoriality in terms of total mutual exclusion and to assume coincidence between the contours of society and state boundaries "the territorial trap."[27] Kären Wigen and Martin Lewis identified the "contiguity-fetish" of political geography as a critical cartographic challenge.[28] While contemporary maps tend to represent spatial units of higher order as "discrete, contiguous blocks," they argued, many sociospatial orders are discontinuous and would be more aptly represented as "lattices, archipelagos, hollow rings, or patchworks." This cartographic convention naturalizes a "jigsaw-puzzle view of the world"[29] in which maps are expected to "show a set of sharply bounded units that fit together with no overlap and no unclaimed territory."[30] Sustained critiques of these graphical conventions can also be found in cartography and geographic information science.[31] It is important to note that the preference for clear, unambiguous modes of visual representation is not unique to cartography but reflects much broader patterns of ambiguity aversion and uncertainty omission among those concerned with visualizing data.[32]

Historians have offered creative propositions for how the Holy Roman Empire's spatial order can be reconceptualized. One particularly imaginative proposal stems from two French historians, Christophe Duhamelle and Falk Bretschneider, who describe the spatial order of the Holy Roman Empire as "fractal."[33] They propose the concept "fractality" as an alternative to the network metaphor because it describes the empire's multiple levels of political organization and their interconnections. As geometric shapes, fractals possess three qualities that make them a fitting metaphor. First, all parts of a fractal, at all scales, are similar to the whole (self-similarity): in the empire, similar institutions and procedures—such as plural forms of territoriality—can be observed at all levels. Second, all parts are connected with each other across scales, making it impossible to make out an unambiguous hierarchy; for example, the lack of a strong central authority led to a circular rather than linear process of enforcement, in which local authorities had to enact the decisions of higher authorities. Third, fractals have no clear center or symmetry: the empire's legal system, for instance, was marked by polycentrism, overlap and competition between courts, and fluidity between levels. A key element of the fractality model is that it considers hierarchy without framing the empire in terms of stacked levels. While Bretschneider and Duhamelle's discussion of fractality is made concrete with a wide range of compelling examples, it is primarily a conceptual rather than visual proposition.

The effort to visualize, rather than just verbally describe, the Holy Roman Empire's geopolitical complexity should be seen as a worthwhile intellectual endeavor in its own right. In this respect, one body of work that deserves particular attention is the regional historical atlas published in Germany—a genre of historical cartography that is remarkable for its detail, diversity, and frequent originality.[34] While many of the maps published in these atlases follow more traditional conventions, readers will also find ingenious attempts at representing shared or disputed dominion, time variance, or pointillist forms of dominion. Some of these attempts push the cartographic medium to its limits, as in Teresa Neumeyer and Robert Winkelbauer's perplexing map of disputed jurisdiction in the *territoria inclausa* of Franconia.[35]

Indeed, it is noteworthy that historians whose work has primarily taken textual form are rediscovering the cartographic representation of the Holy Roman Empire's social and political history as a meaningful avenue of research.[36] Bretschneider and Duhamelle recently co-initiated a Franco-German research group that has begun to develop new ways of visualizing the patterns of "fragmentation, entanglement, plurality, and competition" that were so characteristic of the Holy Roman Empire.[37] As an alternative to patchwork maps like the one above (figure 10.1)—premised on a conception of early modern political space as fragmented but unambiguous—the visualizations created in this project are pointillist rather than polygonal and also engage networks and other noncartographic formats. At Oxford, a group led by Peter Wilson is creating new maps of the Thirty Years' War that go beyond military history's preoccupation with battles or demographic maps of population decline. Instead, their declared aim is to map interactions between "physical and political geography, settlement patterns, communications infrastructure, troop movements and concentrations,

with seasonal variations over time and space."[38]

In my own work at Stanford's Spatial History Project, I questioned the use of polygons in the representation of polities in the Holy Roman Empire, suggesting point-based representations as an alternative.[39] Maps structured by contiguous polygons suggest continuous distribution of phenomena that were in fact discontinuous and reinforce ideas of monocratic rule that have been called into question. The appeal of pointillist mapping has a much longer history and spans different disciplines, as William Rankin has argued. In my case, point-based maps of toll and customs stations allowed me to gain new insights into the spatial patterns of authorities' interference with human mobility, which was unevenly distributed along roads rather than at the outer perimeters of a polity.[40] The following sections aim to further these reflections by discussing ways of visually representing one of the most common yet multiform specificities of the empire's political order: dominion that was shared, contested, or otherwise plural rather than exclusive.

Density and Dilution

One particularly salient characteristic of many retrospective political maps of the early modern German lands, at all scales, is the density of distinct claims to dominion. Familiar to both academic and lay audiences, "patchwork" maps of the Holy Roman Empire are regularly "discovered" and ridiculed on social media. Such choropleth maps are effective at highlighting the degree to which competing claims of dominion structured the empire's landscape of power, even at the largest of scales.[41] The close juxtaposition of shapes encoded with different hues suggests a space saturated with power, a dense mosaic of authority that assigns political power a prominent place in the cartographic composition. One drawback of representing political fragmentation in this form is that it elevates the importance of political power in the map's visual hierarchy, making it easy to forget that the populations inhabiting these spaces often experienced political density as a weakening rather than as a strengthening of authority.[42]

In *The Art of Not Being Governed*, the anthropologist James Scott offered a systematic reflection on how the paradoxical effects of political density could be made visible. Scott developed a series of metaphors and schematic representations to describe the overlap and waning political influence of Southeast Asian polities at their peripheries. Among these were mandalas—circles in which power concentrates at the center and gradually fades into absence at the outer perimeter. Areas of "dual sovereignty"[43] that fell in the ambit of more than one overlord could easily become cases of "mutually cancelling," weakened sovereignty as local chiefs and their subjects enjoyed greater autonomy. The populations in those buffer zones could face competing exactions or punitive raids, but many hill peoples and petty chieftaincies managed to strategically manipulate competition between stronger powers.[44]

Without playing down the differences between Southeast Asia and Central Europe, the idea that political density can lead to "mutually cancelling"

claims of dominion are easily transferred to the early modern German lands, where local populations often found productive ways of turning competition between authorities to their advantage. A prominent example is the practice of *Auslaufen*. When religious minorities lived in areas where dissenting worship was not tolerated (for example, Protestant subjects in a re-Catholicized principality), dissenters often attended services in neighboring communities where their form of worship was tolerated.[45] This was a particularly common practice in the politically densest parts of the Holy Roman Empire, where large parts of the population lived in close proximity to political boundaries. Scott's mandalas offer a good model to visualize the proximity of competing places of worship and their dilutive effects on political authority in matters of religion.

Another domain in which overlapping claims of dominion regularly led to dilution of monocratic power was the exercise of mercy. Petitioning was a ubiquitous practice in old-regime societies, one that fulfilled important functions of conflict resolution, political participation, and communication between authorities and subjects.[46] Acts of mercy or clemency could also neutralize or short-circuit the rulings and decisions of rival or subordinate rulers. In the late fifteenth century, for example, the Swiss Confederacy granted letters of safe conduct to people sentenced by Konstanz's courts. By providing defendants with legal immunity, the confederacy thwarted the execution of judgments and made Konstanz's jurisdiction dependent on its goodwill.[47] The French historian Léonard Dauphant attempted to reconstruct the spatial patterns of competing acts of mercy in late medieval France.[48] Evaluating the records of tens of thousands of remission and appeal procedures, Dauphant compared the spatial distribution of instances of mercy accorded by the kings of France and different princes, including a map of the "rivalry of pardons" between the king of France and the dukes of Burgundy (and their Habsburg successors) in Artois and Flanders. His findings complicated established views of the region's political and linguistic geography.

Mercy played a similarly important role in the Holy Roman Empire, where subjects regularly appealed to the clemency of territorial rulers at all levels, including local authorities as well as the emperor and imperial institutions.[49] Recent research has highlighted the extent to which subjects petitioned not just their immediate authorities, but also the emperor.[50] The archival records of the emperor's Aulic Council include a large number of petitions from across and outside the empire that have recently been catalogued.[51] Subjects petitioned the emperor for a wide range of reasons: to obtain privileges or money, for support in disputes around inheritance or debt, or for protection in matters of criminal justice. In two-thirds of the cases, the decision fell in favor of the petitioners.

Figures 10.2 and 10.3 show my attempts to plot the places of residence of more than 1,300 subjects who petitioned the emperor in the years between 1576 and 1612. Figure 10.2, which plots this data against the outer bounds of the Holy Roman Empire, shows that many petitioners resided in Prague or in the imperial cities in the south, although petitions from northern cities or Saxony were also common. The same data can be visualized in continuous clusters of varying density, as in figure 10.3, which highlights the areas where most of the petitioners were concentrated. The light clusters indicate areas in which subjects were particularly prone to petition the emperor directly when they came into conflict with local authorities. The relationship of background and foreground is inverted: darker shades indicate areas where imperial acts of mercy or grace where less likely to limit the actions of local authorities and territorial rulers. In contrast, the lighter areas indicate "holes" in this fabric of power: subjects in these areas did not hesitate to appeal to the empire's highest authority when they came into conflict with local power holders. The fluid aesthetics generated by the smoothing algorithm reinforces the dilution metaphor.

This representation of the "geography of grace" is, of course, highly schematic. The archival evidence it is based on is mediated and incomplete, not least because petitions were often produced orally. Population density is not taken into account. Neither were all petitions successful; nor does the existence of a record tell us much about its concrete effects in local power dynamics. Nevertheless, I hope this map of clemency illustrates what a visual argument can look like if competing claims of power are framed in terms not of density but of dilution: where the visual language of fragmentation presents us with contiguous, self-contained units, the visual metaphor of dilution frames the same evidence in terms of indeterminacy and contamination.

FIGURE 10.2. Places of residence of subjects petitioning to the Aulic Council between 1576 and 1612. Many petitioners resided in Prague and in imperial cities. The map does not show places of residence that could not be located or disambiguated and places that fall outside the map extent. Map by Luca Scholz. Data sources: DFG/FWF-Projekt Untertanensuppliken am Reichshofrat Kaiser Rudolfs II (database), Austrian Academy of Sciences (HistoGIS).

FIGURE 10.3. The density of petitions framed as dilution. Lighter areas appear as "holes" in the Holy Roman Empire's fabric of power. Map by Luca Scholz. Data sources: "DFG/FWF-Projekt Untertanensuppliken am Reichshofrat Kaiser Rudolfs II" (database); Austrian Academy of Sciences (HistoGIS).

Infrastructure and Orientation

"Patchwork" maps not only suggest that political power in the Holy Roman Empire was particularly dense; they also represent its shape as areal. In practice, we know that the spatial forms of political power in the empire were much more variegated. Lines, often joined into networks, are one type of geometry that historians and their collaborators regularly suggest as an alternative to areal notions of territoriality. The economic historian Monica Smith, for example, has criticized the representation of ancient states as territorial blobs as a "convenient fiction"—one that overstates the capacity of central governments to control large territories and exaggerates the importance of boundaries that were in practice "porous, permeable, flexible, and selectively defended." Instead, Smith calls for visualizing ancient states as networks of resource acquisition. She prefers the network model to represent ancient polities because it emphasizes the importance of state-level control over road infrastructure and nodal points—more cost effective than the control of territory—and because it better reflects the role of kinship and alliances. Examples for ancient states that would be better described by network models include the linear polity of the Phoenicians around the Mediterranean and nodal polities such as the nomadic dynasties of Eastern Africa or the empires of Central Asia. Notably, as a geopolitical metaphor, networks emphasize cooperation, while contiguous polygons frame the coexistence of polities in terms of separation and exclusion rather than cooperation and association.[52]

Network maps are particularly appropriate to visualizing competing claims of dominion over the Holy Roman Empire's roads and rivers, as well as those who moved on them. Conflicts concerning the governance of the mobility of goods and people were often framed as matters of safe conduct (*Geleit*).[53] Safe conduct could take different forms and meant different things for different people: it could describe the act of escorting a traveler, letters issued to debtors or felons, a duty to pay transit duties, or an abstract concept. In early modernity, it evolved from a primarily protective service into a broad set of administrative and political instruments, symbolic practices, and a conceptual framework that authorities and mobile populations used to negotiate the protection, promotion, and regulation of different forms of interpolity mobility.

A key spatial characteristic of safe conduct was that the right was anchored to stretches of roads and rivers that did not always overlap with a ruler's other claims of dominion. Safe conduct is thus a good illustration for how the spatial organization of many imperial estates followed a "seigneurial"[54] rather than a "territorial logic," which could resemble an "aggregation of titles of ownership"[55] more than homogeneous, bounded units. The lack of spatial cohesion complicated the exercise of rights like safe conduct—sometimes to the point of making it impossible—and meant that safe-conduct officials regularly exercised their duties in foreign territory.[56] It also made safe conduct an important tool of territorial control and expansion. The Electors Palatine, for example, used their control of safe conduct in the

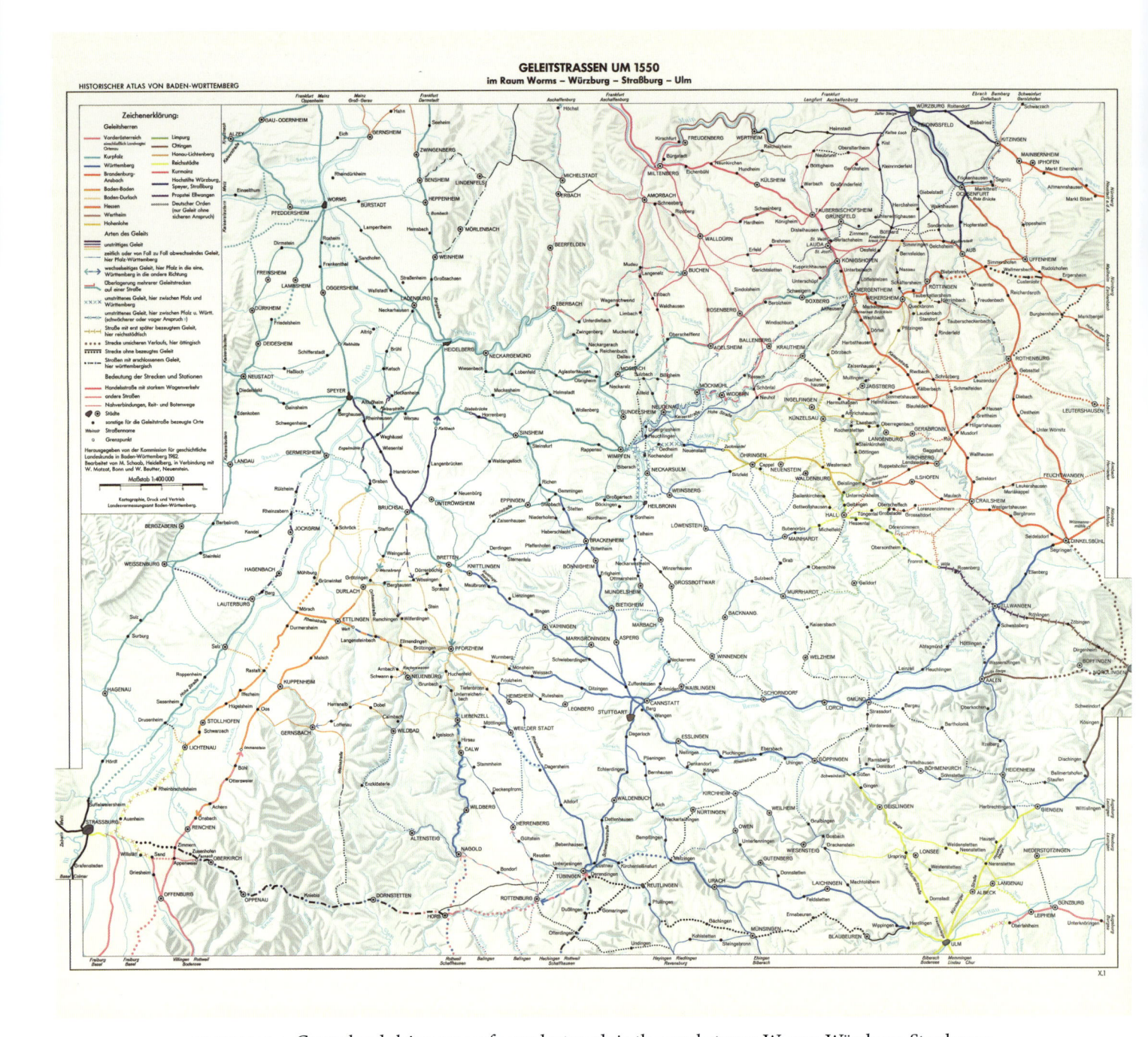

FIGURE 10.4. Control and claims over safe-conduct roads in the area between Worms, Würzburg, Strasbourg, and Ulm around 1550. Meinrad Schaab, Wilhelm Matzat, and Wilfried Beutter, *Geleitstrassen um 1550 im Raum Worms-Würzburg-Straßburg-Ulm*, in *Historischer Atlas von Baden-Württemberg*, ed. Kommission für Geschichtliche Landeskunde in Baden-Württemberg and Landesvermessungsamt Baden-Württemberg (Stuttgart: Kommission für Geschichtliche Landeskunde in Baden-Württemberg, 1982), 10:1.

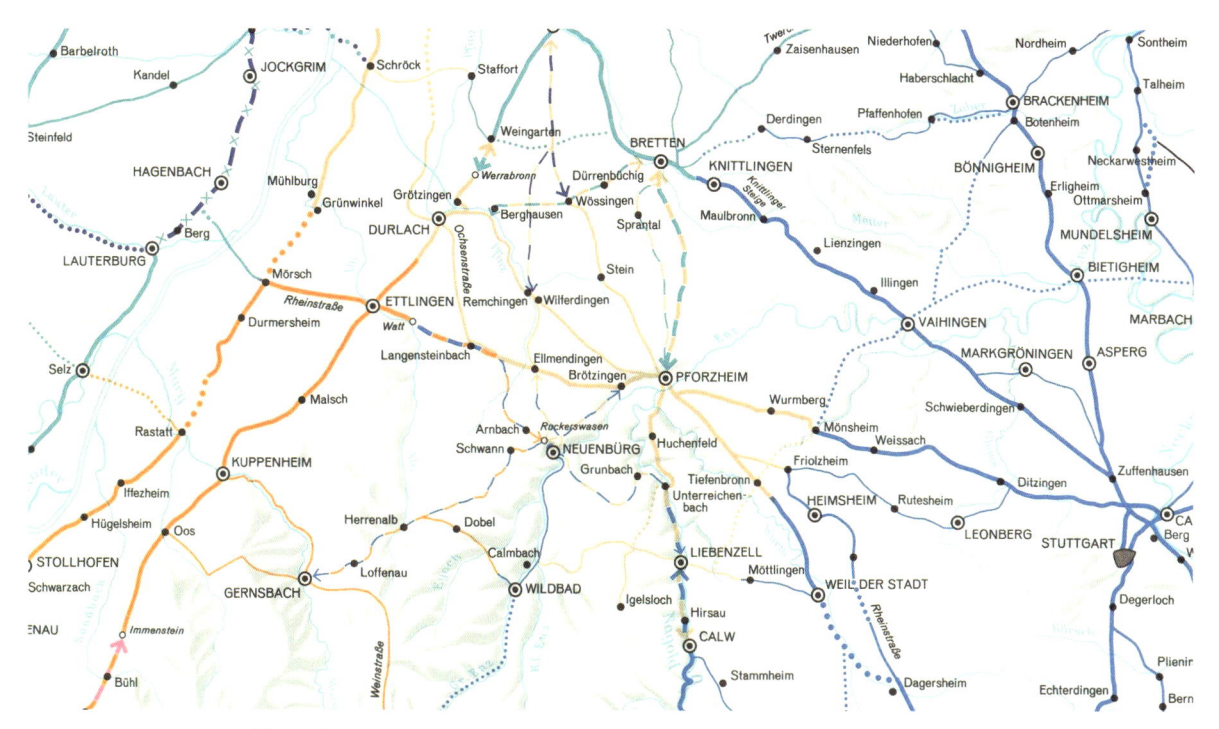

FIGURE 10.5. Detail from figure 10.4. Where competing claims of safe-conduct overlapped, neighboring authorities found a range of solutions, such as dividing its exercise by orientation of travel.

lands of less powerful neighbors as a means of extending their influence.[57] The city of Bremen used its convoy ship on the Lower Weser to impose its claims of dominion and protect its economic interests.[58] Safe conduct also played an important role in efforts to channel the flows of merchants, carters, and their goods through a territory and its toll stations—for example, by criminalizing the use of certain roads and designating others as the only authorized "ordinary" routes.[59]

The map in figure 10.4 represents the most important roads on which rulers in the southwestern parts of the Holy Roman Empire held or claimed rights of safe conduct in one of the empire's most politically dense regions. It was designed by historians and a geographer and published in a regional historical atlas. This map is noteworthy because it stands apart from how political geography in the empire is conventionally represented in historical maps

and atlases. Instead of framing territorial claims as contiguous, enclosed polygons over which power is exercised homogenously, it gives them a linear shape. As a visual argument, the map thus emphasizes the importance of road infrastructure for the control of territory. Maps like the one above invite readers "to think about the state less as a unitary container of populations and more as a more variegated, incomplete, and dispersed network of order and control."[60] The graphical model emphasizes the importance of infrastructure and gives visual expression to the assumption that states are "spatially malleable entities" whose power is concentrated to different degrees in different settings.[61]

The map is also of interest for the rich graphical language employed to encode the ways these polities shared, alternated, claimed, or disputed their control of road infrastructure (figure 10.5). While competing claims over the same roads and rivers regularly led to confrontation (most such conflicts being tamed by legal protocol, but some escalating into physical violence), many polities found arrangements that allowed the exercise of safe conduct while maintaining each side's claims. For instance, some rulers appointed shared officials who performed the safe-conduct duties over contested routes in the name of both parties. Where several safe-conduct officials were appointed for the same roads, they sometimes escorted travelers in opposite directions, sparing the escorts lengthy waits in the countryside and the difficult negotiation of a common boundary. This is one example in which maps can account for directionality and isotropy and challenge the notions of homogeneous sovereignty that Franck Billé critiques in chapter 7 of this volume. The safe-conduct map under discussion uses colored arrows to make visible this form of shared dominion, which sits uneasily with the notions of fragmentation, exclusion, or separation; this was, rather, dominion divided by orientation.

Oscillation

Another circumstance that complicates efforts to represent the Holy Roman Empire on maps is the frequency with which its political geography varied in time. Frequent partitions, acquisitions, or swaps of land and titles made the empire's territories a highly dynamic political landscape, a close-up of which would have revealed "something akin to a mass of amoebae, constantly changing shape," as Joachim Whaley put it.[62] Yet making time variance visible is not just a matter of cartographic accuracy or detail;

it can offer important insights into the forces that shaped these processes. The history of safe conduct, for example, shows that authorities were often forced to move their toll stations to follow shifting traffic flows.[63] Such decisions illustrate the extent to which patterns of mobility could influence the spatial make-up of early modern polities and their modes of accumulating power and resources: flows of goods and people shaped the geography of these polities just as much as the polities could shape flows in turn.

Time variance was also constitutive to some forms of the *condominium*, a particularly formalized and long-lived model of joint dominion. In Europe, condominia are documented at least since the sixth century, and many continued to exist well into the nineteenth century.[64] These joint dominions came about for different reasons—through succession, as in the case of *Ganerbschaft*, or as results of regional political arrangements—but their purpose was almost invariably that of balancing the interests of competing polities, rulers, estates, or other groups. Condominia could be of considerable size and administered directly (whether immediately or through deputies) or through jointly appointed officials. The older historiography has often emphasized the administrative, jurisdictional, or religious conflict and disfunction engendered by these shared power arrangements. In contrast, many early modern publicists had a more nuanced vision of these forms of political organization, an attitude that is also echoed by more recent scholarship emphasizing that condominia were long-lived and functional models for balancing competing claims of dominion.

FIGURE 10.6. In the mid-eighteenth century, high jurisdiction over the entire Frais district (*enclosed by the dotted and dashed lines*) was shared as a condominium between the Electorate of Bavaria (*bottom*) and the Kingdom of Bohemia (*top*). High jurisdiction over the *Wechselfrais* (*enclosed by only the dotted line*) switched annually between Bavaria and Bohemia. Heribert Sturm, *Tirschenreuth* (Historischer Atlas von Bayern, Altbayern I/21, Kommission für bayerische Landesgeschichte, 1970), 141.

One such example is the Frais condominium, a small district in the borderland between Bohemia and Bavaria. The Frais district was shared by the Kingdom of Bohemia (for the city of Eger, now Cheb) and the Electors Palatine, later the Electorate of Bavaria (both for the Waldsassen Abbey). It was established in 1591, after protracted negotiations following the murder of an artisan.[65] The new treaty established a yearly switching condominium for a part of the Frais district, the so-called *Wechselfrais*. This

condominium was initially envisaged as a provisional solution, but the two parties were unable to agree on how to divide up the district. As a consequence, the provisional treaty remained in place until the mid-nineteenth century, even as Waldsassen was incorporated into the Electorate of Bavaria in 1628 and Eger gradually lost its independence, which meant that the *Wechselfrais* eventually transformed into a condominium between Bavaria and Austria (figure 10.6).

The population of the *Wechselfrais* lived in four

"mixed" towns with subjects of both parties—Neualbenreuth, Altalbenreuth, Gosel, and Querenbach—and five "unmixed" towns whose inhabitants were subject to Waldsassen: Hardeck, Schachten, Boden, Mugl, and Maiersreuth. High jurisdiction over these settlements switched every year on July 29. In the first years, the annual switch of dominion was formally announced on the marketplace in Neualbenreuth. The annually alternating claims of dominion did not eliminate all differences between the two parties; however, despite frequent and protracted disputes, the treaty of 1591 provided a lasting legal foundation that neither party ever seriously called into question.

Existing historical maps of the Frais district represent the *Wechselfrais* with dotted lines, as in figure 10.6, or do not differentiate the temporal condominium from the remaining condominium at all.[66] Dotted and hatched textures are well-established visual variables for representing condominia in historical atlases of the early modern German lands. Figure 10.7 is an attempt at representing the *Wechselfrais* as qualitatively different from other condominia: because high jurisdiction over this space alternated every year, there was no moment in time at which it was exercised jointly. The composition employs a simple but well-established cartographic technique for visualizing time variance: small multiples. Like snapshots in time, maps are arranged in a sequence to show how high jurisdiction over the *Wechselfrais* changed—or, rather, oscillated—over time.[67] It is a simple, and possibly not entirely satisfactory, at-tempt to visually represent this condominium not as a shared space governed and adjudicated jointly by the two regional powers, but as a space in which jurisdiction was shared diachronically yet was exclusive at any given moment.

Conclusion

This chapter has proposed visual strategies to address the cartographic lag that separates textual and visual arguments on the political geography of premodern German lands. The representation of the Holy Roman Empire as a patchwork of contiguous, self-contained, and fragmented polities has been called into question on practical as well as conceptual grounds. The recent historiography on the empire has revised many of the assumptions inherent in the metaphors of contiguity, self-containment, and fragmentation and proposed alternative ways of conceptualizing the empire's spatial and political order. While most of these interventions have been primarily conceptual rather than cartographic, recent years have seen renewed interest in cartographic representation as a form of historical argument. I have proposed three ways in which nonexclusive forms of dominion can be accounted for on maps: by reframing political density as dilution; by emphasizing the importance and idiosyncrasies of infrastructure, including distinctive forms of shared dominion such as division by orientation; and by considering diachronic modes of power sharing as constitutive of the empire's political geography.

FIGURE 10.7. High jurisdiction over the *Wechselfrais* switched annually between Bavaria (*white*) and Bohemia (*black*). This arrangement in small multiples is an attempt at conveying the circumstance that while high jurisdiction over this area was shared, it remained exclusive at all times. Map by Luca Scholz. Data source: Heribert Sturm, *Tirschenreuth* (Historischer Atlas von Bayern, Altbayern I/21, Kommission für bayerische Landesgeschichte, 1970).

As other chapters in this volume show, reconsiderations of sovereignty couched in the language of exception should be questioned because they normalize monocratic, unambiguous, and self-contained forms of dominion and frame nonexclusive forms of political power as abnormal and deviant. This is as true for the concepts and terminology employed to describe them as for the visual language of retrospective cartography. Franck Billé argues in chapter 7 that bolder and more imaginative conceptual and cartographic responses are called for than "hyphenated qualifiers of sovereignty."[68] Evolving graphical conventions can play an important role in conveying shifting scientific and scholarly paradigms, and the maps presented in this chapter are experimental propositions aimed at furthering this line of thinking.[69] If carefully designed, retrospective cartography—as this chapter has tried to show—can play an important role in retraining the eye to see through and beyond the sovereign map.

NOTES

Introduction

1. On the ties between memory and place in a different Native community, see Keith H. Basso, *Wisdom Sits in Places: Landscape and Language among the Western Apache* (University of New Mexico Press, 1996).

2. Henry R. Hermann, "Traits of Dominant Animals," in *Dominance and Aggression in Humans and Other Animals* (Elsevier, 2017). The classic study in geography is Robert Sack, *Human Territoriality: Its Theory and History* (Cambridge University Press, 1986).

3. Max Weber, "The Profession and Vocation of Politics," in *Political Writings*, ed. Peter Lassman and Ronald Speirs (1919; Cambridge University Press, 1994), 309–69, quotation on 310–11; italics in the original.

4. John Gillis, *Islands of the Mind: How the Human Imagination Created the Atlantic World* (Palgrave Macmillan, 2004).

5. Landmarks include Clarence Glacken, *Traces on the Rhodian Shore: Nature and Culture in Western Thought from Ancient Times to the End of the Eighteenth Century* (University of California Press, 1976); Edward Said, *Orientalism: Western Conceptions of the Orient* (Vintage, 1979); and Thongchai Winichakul, *Siam Mapped: A History of the Geo-body of a Nation* (University of Hawai'i Press, 1997). Since the mid-1990s, the literature has burgeoned. For representative works, see Martin Lewis and Kären Wigen, *The Myth of Continents: A Critique of Metageography* (University of California Press, 1997); Margaret Cohen, *The Novel and the Sea* (Princeton University Press, 2012); Sumathi Ramaswamy, *The Goddess and the Nation: Mapping Mother India* (Duke University Press, 2010); and Geoffrey Gunn, *Imagined Geographies: The Maritime Silk Roads in World History, 100–1800*

(Hong Kong University Press, 2021).

6. Stephen Krasner in fact distinguishes four separate components of sovereignty: *international legal sovereignty* ("the practices associated with mutual recognition"); *Westphalian sovereignty* ("the exclusion of external actors from authority structures within a given territory"); *domestic sovereignty* ("the ability of public authorities to exercise effective control within the borders of their own polity"); and *interdependent sovereignty* ("the ability of public authorities to regulate the flow of information, ideas, goods, people, pollutants, or capital across the borders of their state"). Stephen Krasner, *Sovereignty: Organized Hypocrisy* (Princeton University Press, 1999), 3–4.

7. Wendy Brown, *Walled States, Waning Sovereignty* (MIT Press, 2014), 48.

8. Brown, *Walled States, Waning Sovereignty*, 49.

9. James Sheehan, "The Problem of Sovereignty in European History," *American Historical Review* 111, no. 1 (2006): 1–15, quotation on 2.

10. Luca Scholz, *Borders and Freedom of Movement in the Holy Roman Empire* (Oxford University Press, 2020), 32.

11. Scholz, *Borders and Freedom of Movement*, 89. To this day, control over the movement of people and goods is often as important as, if not more important than, control over borders. See Peer Schouten, *Roadblock Politics: The Origins of Violence in Central Africa* (Cambridge University Press, 2022).

12. Fabian Drixler, William D. Fleming, and Robert George Wheeler, *Samurai and the Culture of Japan's Great Peace* (Yale Peabody Museum, 2015), 23.

13. Gregory Smits, *Maritime Ryukyu, 1050–1650* (University of Hawaiʻi Press, 2019).

14. Ronald P. Toby, "Mapping the Margins: The Ragged Edges of State and Nation," chap. 2 of *Engaging the Other: "Japan" and Its Alter-Egos, 1550–1850* (Brill, 2019), 25–73. For more on this theme, see David Howell, *Geographies of Identity in Nineteenth-Century Japan* (University of California Press, 2005); and Bruce L. Batten, *To the Ends of Japan: Premodern Frontiers, Boundaries, and Interactions* (University of Hawaiʻi Press, 2003).

15. Mark Ravina, "Rethinking Historical Maps for the 21st Century: A Quantitative Perspective on Japan's *kuniezu*" (talk delivered at Nichibunken [International Research Center for Japanese Studies], June 9, 2023).

16. A number of political scientists and other scholars have noted this problem. See, for example, Robert Jackson, *Quasi-states: Sovereignty, International Relations, and the Third World* (Cambridge University Press, 1990).

17. Denis Wood, *Rethinking the Power of Maps* (Guilford, 2010), 33.

18. Lauren Benton, *A Search for Sovereignty: Law and Geography in European Empires, 1400–1900* (Cambridge University Press, 2009), 2.

19. According to Google Books Ngram Viewer, the tipping point occurred in roughly 1890.

20. See Melissa M. Lee, Nan Zhang, and Tilmann Herchenrödere, "From Pluribus to Unum? The Civil War and Imagined Sovereignty in Nineteenth-Century America," *American Political Science Review* 118, no. 1 (February 2024): 127–43. Native nations effectively constitute their own sovereign exclaves in the national fabric, overlaying the extraordinary legal patchwork created by the seismic regulatory differences between the fifty states.

21. William Rankin, *Radical Cartography: Visual Argument in the Age of Data* (Viking Books, forthcoming 2025; cited with permission).

22. Fabian Drixler's team at Yale is at work on a Digital Tokugawa Atlas, described at https://dtl.macmillan.yale.edu /digital-atlas-tokugawa-japan. At University of Texas at Austin, Mark Ravina directs the more open-ended JapanLab, https://www.utjapanlab.com/, which supports work on video games and podcasts as well as timelines and maps. Ravina's own promising effort to remap the Tokugawa order—taking into account the ambiguities of borders and the fragmentation of domains—is still underway, with preliminary results published in Mark Ravina, "Algorithmic Maps and the Political Geography of Early-Modern Japan," *Journal of Cultural Analytics* 8, no. 3 (August 29, 2023), https://doi.org/10.22148/001c.84860. For more experiments in digital historical cartography, see the archived web pages of the Stanford Spatial History Project, which was active from 2005 to 2022: http://web.stanford.edu/group /spatialhistory/static/.

23. It is worth noting here a conceptual overlap between the essays of Bol and Munkh-Erdene. Both chapters explore the contrived political genealogies of China, which typically trace the history of the modern state back from the present through the Qing, the Ming, the Song, the Han, and often on to the Shang and prehistory. Both Munkh-Erdene and Bol challenge that kind of linear narrative. The first map Bol discusses is a product of the Qi dynasty, an enemy of the Song, while the Qing was, of course, ethnically Manchu rather than "Han Chinese." While focusing on questions of mapping, in other words, both these essays also engage questions of "Chinese" political identity. I am grateful to an anonymous reader for pointing out this shared concern.

24. Daniel Immerwahr, "Zoning Out," review of Quinn Slobodian, *Crack-Up Capitalism: Market Radicals and the Dream of a World Without Democracy*, *New York Review of Books*, November 23, 2023, 22–25, quotation on 22.

Chapter 1

1. The literature on the incorporation of Siberia is extensive. Among others, see Iu. G. Akimov, "Political Claims, an Extensible Name, and a Divine Mission: Ideology of Russian Expansion in Siberia," *Journal of Early Modern History* 25 (2021): 277–99; Valerie Kivelson, *Cartographies of Tsardom: The Land and Its Meanings in Seventeenth-Century Russia* (Cornell University Press, 2006); Yuri Slezkine, *Arctic Mirrors: Russia and the*

Small Peoples of the North (Cornell University Press, 1994); A. S. Zuev, P. S. Ignatkin, and V. A. Slugina, *Pod sen' dvuglavogo orla: inkorporatsiia narodov Sibiri v Rossiiskoe gosudarstvo v kontse XVI–nachale XVIII v.* (IPTS Novosibirsk State University, 2017); and A. S. Zuev and V. A. Slugina, "Legitimatsiia vlasti rossiiskogo monarkha nad Sibir'iu i ee narodami v kontse XVI–nachale XVIII v.," *Vestnik RUDN* (Rossiiskii universitet druzhby narodov), Istoriia Rossii (series), vol. 20, no. 3 (2021): 340–52.

2. S. U. Remezov, Chertezhnaia kniga, Rossiiskaia gosudarstvennaia biblioteka, Moscow, Rukopisnyi otdel, coll. 256, Rumiantsev Collection, no. 346; published in a facsimile edition as "Chertezhnaia kniga Sibiri, sostavlennaia tobol'skim synom boiarskim S. Remezovym v 1701 godu," 2 vols. (FGUP, PKO Kartografiia, 2003), available online at https://viewer.rusneb.ru/ru/000199_000009_004274925; Semen Ul'ianovich Remezov, 1642–ca. 1720. *Khorograficheskaya kniga* [cartographical sketchbook of Siberia]. MS Russ 72 (6). Houghton Library, Harvard University, Cambridge, MA. Published in facsimile: Semen Ul'ianovich Remezov, *Khorograficheskaia chertezhnaia kniga Siberi S. U. Remezova (1697–1711)* (Fond "Vozrozhdenie Tobol'ska," 2011). Available online at https://nrs.lib.harvard.edu/urn-3:fhcl.hough:4435676. "Sluzhebnaia chertezhnaia kniga Remezova," Rossiiskaia natsional'naia biblioteka, St. Petersburg, Ermitazhnoe sobranie, no. 237. Published in facsimile: *Sluzhebnaia chertezhnaia kniga* (Fond "Vozrozhdenie Tobol'ska," 2006). Available online at https://viewer.rusneb.ru/ru/000200_000018_NLR_%D0%9E%D0%A0%2040AC-6CFA%20D8%3%2047B2%20B472%207ABF833BF9D4?page=1.

3. Leonid A. Gol'denberg, *Izograf zemli sibirskoi* (Magadanskoe knizhnoe izdatel'stvo, 1991). A posthumous piece by Gol'denberg, edited by A. V. Postnikov, is available in English: "Russian Cartography to ca. 1700," in *Cartography in the European Renaissance*, vol. 3 of *The History of Cartography*, ed. David Woodward (University of Chicago Press, 2007), 1852–903.

4. A. V. Efimov, ed., *Atlas geograficheskikh otkrytii v Sibiri i v severo-zapadnoi Amerike XVII-XVIII vekov* (Nauka, 1964), xvi, 31, 34.

5. A. V. Psianchin, "Istoriia etnicheskoi kartografii v Rossii (do 30-x gg. XX v.)" (dissertation abstract [*avtoreferat*], Institute for the History of Natural Science and Technology, Moscow, 2004), 18.

6. My thanks to Kären Wigen for sharing her perceptive thoughts on this map, which I draw on here. Email communication, June 6, 2022.

7. Valerie Kivelson, "Angels in Tobolsk: Celestial Topography and Visionary Administration in Late Muscovite Siberia," *Harvard Ukrainian Studies* 28, no. 1/4 (2006): 543–56.

8. John H. Appleby, "Mapping Russia: Farquharson, Delisle and The Royal Society," *Notes and Records of the Royal Society of London* 55, no. 2 (2002): 191–204, quotation on 192.

9. On the history of establishing and mapping the Russian–Chinese border, see A. V. Postnikov, *Istoriia geograficheskogo izucheniia i kartografirovaniia Sibiri i Dal'nego Vostoka v XVII-nachale XX veka v sviazi s formirovaniem russko-kitaiskoi granitsy* (URSS, 2015).

10. Yuri Slezkine, "The USSR as a Communal Apartment, or How a Socialist State Promoted Ethnic Particularism," *Slavic Review* 53, no. 2 (1994): 414–52.

11. See Gregory Afinogenov, *Spies and Scholars: Chinese Secrets and Imperial Russia's Quest for World Power* (Belknap, 2020); Pamela Kyle Crossley, *A Translucent Mirror: History and Identity in Qing Imperial Ideology* (University of California Press, 1999); James Meador, "Cossacks into Manchus: Transfrontier Intermediaries in Inner Northeast Asia," *Ab Imperio*, no. 3 (2021): 75–110; Sören Urbansky, *Beyond the Steppe Frontier: A History of the Sino-Russian Border* (Princeton University Press, 2020); and Eric Widmer, *The Russian Ecclesiastical Mission in Peking during the Eighteenth Century* (Harvard University Asia Center, 1976).

12. Meador, "Cossacks into Manchus"; quotations taken from a draft version of the article.

13. Meador; quotations taken from a draft version of the article.

14. "The two empires had gradually groped toward each other across vast underpopulated spaces in the seventeenth century." Peter C. Perdue, *China Marches West: The Qing Conquest of Central Eurasia* (Belknap Press of Harvard University Press,

2005), 163. See also Stephen Kotkin, "Mongol Commonwealth? Exchange and Governance across the Post-Mongol Space," *Kritika: Explorations in Russian and Eurasian History* 8, no. 3 (2007):487–531; and Peter C. Perdue, "Boundaries, Maps, and Movement: Chinese, Russian, and Mongolian Empires in Early Modern Central Eurasia," *International History Review* 20, no. 2 (1998): 263–86.

15. Widmer, *Russian Ecclesiastical Mission*, 10; also quoted in Meador, "Cossacks into Manchus," 81.

16. Gregory Afinogenov gives a lively view of the individual intermediaries who facilitated or blocked on-the-ground interactions before and after the founding of the Beijing Orthodox Mission. Gregory Afinogenov, *Spies and Scholars: Chinese Secrets and Imperial Russia's Quest for World Power* (Belknap and Harvard University Press), 2020. See also Mark Mancall, *Russia and China: Their Diplomatic Relations to 1728* (Harvard University Press, 1971).

17. Remezov, Chorographic Sketchbook, chap. 54, fol. 147.

18. *Chertezhnaia kniga,* "*Chertezh zemli Nerchinskago goroda*," map 19, fol. 40.

19. Chorographic Sketchbook, chap. 54, fol. 148.

20. Kicengge, "The Illusion of the Nerchinsk Treaty Boundary-stone: The Map of the Amur Region in Manchu," *National Palace Museum Research Quarterly* 29, no. 1 (2011): 147–236 (English abstract, 60–61). Peter Perdue says the steles were set, and includes an illustration showing them at imposing scale. Perdue, *China Marches West*, 169–70.

21. Urbansky, *Beyond the Steppe Frontier*, 20.

22. Peter Sahlins, *Boundaries: The Making of France and Spain in the Pyrenees* (University of California Press, 1991), 37. See also J. R. Hale, *Civilization of Europe in the Renaissance* (Atheneum, 1994), 34–35; and J. A. Akerman, "Cartography and the Emergence of Territorial States," *Proceedings of The Tenth Annual Meeting of the Western Society for French History*, ed. John F. Sweets (University of Kansas, 1984), 84–93.

23. Nicolas de Fer, *Les Etats du Czar ou Empereur Des Russes en Europe et en Asie, . . .* (Paris, 1722). Available at https://www.raremaps.com/gallery/detail/67129/les-etats-du-czar-ou-empereur-des-russes-en-europe-et-en-asi-de-fer.

24. See https://upload.wikimedia.org/wikipedia/commons/a/a5/1730_map_of_Russia_and_Siberia_by_Strahlenberg.jpg. Efimov notes that the Chaplin map of the region from Tobolsk to Chukotka drawn in connection with the first Bering expedition in 1729/30 was embellished with an imperial Russian double-headed eagle when it was first published by Jean-Baptiste Du Halde in *Description Géographique, Historique, Chronologique, Politique et Physique de l'Empire de la Chine et de la Tartarie Chinoisie* (Paris: P.G. Le Mercier, 1735), vol. 4 of 4. No such imperial imprint appeared on the original drawing. Efimov, *Atlas geograficheskikh otkrytii*, no. 65: 45.

25. Efimov says the map was composed in 1722 on the basis of maps by Nicolaas Witsen and Eberhard Ides. See Efimov, *Atlas geograficheskikh otkrytii*, 41. https://disk.yandex.com/a/VvkGdFvh3ZQRWS/5b525a0a4807a600f8a4343f Or https://www.abebooks.com/maps/Russian-Empire-Siberia-Kamchatka-Ukraine-1750/30650402706/bd#&gid=1&pid=1.

26. I. K. Kirilov, *General'naia karta o Rossiiskoi imperii*, 1734. Available at https://upload.wikimedia.org/wikipedia/commons/6/68/Imperii_Russici_tabula_generalis_opera_et_studio_Ioannis_Kyrilow%2C_1734.jpg.

27. Remezov, *Chertezh Iakutskogo goroda*, in *Chertezhnaia kniga*, map 17, fol. 36.

28. Remezov, map 17, fol. 36.

29. Beiton, quoted in Efimov, *Atlas geograficheskikh otkrytii*, 39. Map reproduced as no. 54. On the allure of fur, see Slezkine, *Arctic Mirrors*.

30. Erika Monahan, "Tents or Towns: The Limits of Sovereignty in the Russian North in the late Seventeenth Century," in *Picturing Russian Empire*, ed. by Valerie Kivelson, Sergei Kozlov, and Joan Neuberger (Oxford University Press, 2024), 89, 90.

31. Remezov, "*Chertezh zemli Turukhanskago zemli (Mangazeia)*," *Chertezhnaia kniga*, map 13, fol. 28.

32. See, for example, Chorographic Sketchbook, chap. 29, fol. 98. See the key to abbreviations in *Chertezhnaia kniga*, fol. 2-3.

33. See Kivelson, *Cartographies of Tsardom*, 171–93.

34. Remezov, Chorographic Sketchbook. chap. 54, inserted at fol. 148. Available at https://iiif.lib.harvard.edu/manifests/view/drs:18273155$181i.

35. Remezov, Chorographic Sketchbook, chap. 54, fol. 150. Available at https://iiif.lib.harvard.edu/manifests/view/drs:18273155$184i.

36. Efimov, *Atlas geograficheskikh otkrytii*, no. 78, p. 52.

Chapter 2

1. Daniel Philpott, *Revolutions in Sovereignty: How Ideas Shaped Modern International Relations* (Princeton University Press, 2001), 17.

2. Henry Maine, *Ancient Law* (John Murray, 1908), 91.

3. Lhamsuren Munkh-Erdene, *The Nomadic Leviathan: A Critique of the Sinocentric Paradigm*, vol. 16 of the Inner Asia Book Series (Brill, 2023), 5, 19, 278–79. See also Elman Service, *Origins of the State and Civilization: The Process of Cultural Evolution* (W. W. Norton, 1975), 3–10; and James Scott, *The Art of Not Being Governed: An Anarchist History of Upland Southeast Asia* (Yale University Press, 2009).

4. Jordan Branch, *The Cartographic State: Maps, Territory, and the Origins of Sovereignty* (Cambridge University Press, 2014), 22. See also John Gerard Ruggie, "Territoriality and Beyond: Problematizing Modernity in International Relations," *International Organization* 47, no. 1 (1993): 149; and Philpott, *Revolutions in Sovereignty*, 17.

5. Andreas Osiander, "Sovereignty, International Relations, and the Westphalian Myth," *International Organization* 55, no. 2 (2001): 251–87.

6. See Robert J. Miller, Jacinta Ruru, Larissa Behrendt, and Tracey Lindberg, *Discovering Indigenous Lands: The Doctrine of Discovery in the English Colonies* (Oxford University Press, 2010). To win sovereignty, thus to become a state, became a matter of recognition by those who held it and it transformed them into a licensing club. The license is now the exclusive prerogative of the UN Security Council.

7. Sharon Korman, *The Right of Conquest: The Acquisition of Territory by Force in International Law and Practice* (Oxford University Press, 1996), 57.

8. Mark F. Lindley, *The Acquisition and Government of Backward Territory in International Law: Being a Treatise on the Law and Practice Relating to Colonial Expansion* (Longmans, Green, 1926), 11. See also Emer de Vattel, *The Law of Nations* (Liberty Fund, 2008), 128–30.

9. John Locke, *Two Treatises of Government*, ed. Peter Laslett (Cambridge University Press, 1988), 339; see also 285–302.

10. Maine, *Ancient Law*, 91, 92.

11. Maine, 90.

12. Maine, 95.

13. Lewis Morgan, *Ancient Society* (Charles H. Kerr, 1877), 6.

14. Morgan, *Ancient Society*, 70, 101.

15. Morgan, 224.

16. Morton Fried, *The Evolution of Political Society: An Essay in Political Anthropology* (Random House, 1967), 45.

17. Morgan, *Ancient Society*, 29.

18. Morgan, 6, 26–27.

19. Frederick Engels, *The Origin of the Family Private Property and the State*, trans. Ernest Untermann (Charles H. Kerr, 1909), 27.

20. Morgan, *Ancient Society*, 4, 151.

21. Alfred Radcliffe-Brown, preface to *African Political Systems*, ed. Meyer Fortes and E. E. Evans-Pritchard (Oxford University Press, 1940), xiv.

22. Meyer Fortes and E. E. Evans-Pritchard, introduction to *African Political Systems*, ed. Meyer Fortes and E. E. Evans-Pritchard (Oxford University Press, 1940), 10.

23. No doubt, Max Weber's emphasis on territory, too, reinforced the people-to-territory transition. However, Weber's emphasis was on "the modern state"; and, after all, it is "the *monopoly of the legitimate use of physical force*" that makes "human community" a state and a piece of land a territory. Max Weber, *From Max Weber: Essays in Sociology*, trans. and ed. H. H. Gerth; C. Wright Mills (Oxford University Press, 1946), 78. See also Max Weber, *Economy and Society: An Outline of Interpretive Sociology*, ed. Guenther Roth and Claus Wittich (University of California Press, 1978), 55. Patrimonial states—the embodiment of Weber's traditional authority that preceded his legal authority (the modern *Rechtsstaat*)—all were territorial. In fact, feudalism was built on a ruler's granting of land in fief, and the

rule was "*nulle terre sans seigneur*." Weber, *Economy and Society*, 257, 226–41, 255–66, 1006–110.

24. Fortes and Evans-Pritchard, introduction to *African Political Systems*, 7.

25. Fortes and Evans-Pritchard, 7.

26. See Andrew Strathern, "Despots and Directors in the New Guinea Highlands," *Man* 1, no. 3 (1966): 356–67; Marshall Sahlins, *Tribesmen*, Foundations of Modern Anthropology (Prentice-Hall, 1968); Robert Carneiro, "A Theory of the Origin of the State," *Science* 169, no. 3947 (1970): 733–38; Paula Brown, "New Guinea: Ecology, Society, and Culture," *Annual Review of Anthropology* 7 (1978): 263–91; Jared Diamond, *Guns, Germs, and Steel: The Fates of Human Societies* (Norton, 1999); and Napoleon Chagnon, *Yąnomamö*, Legacy 6th ed. (Cengage Learning, 2013).

27. Fried, *Evolution of Political Society*, 45–49.

28. Fried, 45–49.

29. Weber, *Economy and Society*, 220.

30. Weber, 220.

31. Weber, 1123.

32. See Lhamsuren Munkh-Erdene, *The Taiji Government and the Rise of the Warrior State: The Formation of the Qing Imperial Constitution*, Inner Asia Book Series, vol. 14 (Brill, 2022).

33. Owen Lattimore, *Manchuria: Cradle of Conflict* (Macmillan Company, 1932), 126.

34. Owen Lattimore, *The Mongols of Manchuria* (George Allen & Unwin, 1935), 146.

35. Lattimore, *Mongols of Manchuria*, 78.

36. Owen Lattimore, *Inner Asian Frontiers of China* (1940; Capitol Publishing and American Geographical Society, 1951), 475.

37. If Bodin's "Tatar king" or "the great king of Tartarie" is none other than the Mongol khan or the Great Khan of the Mongol Empire, for him the only "absolute and sovereign" power was the "great power" of the great king of Tartarie. Jean Bodin, *The Six Bookes of a Commonweale*, trans. Richard Knolles (Impensis G. Bishop, 1606), 88–89. See also Jean Bodin, *On Sovereignty: Four Chapters from the Six Books of the Commonwealth*. ed. and trans. Julian H. Franklin (Cambridge University Press, 1992), 8.

38. Owen Lattimore, "On the Wickedness of Being Nomads," in *Studies in Frontier History: Collected Papers 1928–1958* (1935; Oxford University Press, 1962), 415–26. See also Sahlins, *Tribesmen*, 6–7; and Scott, *Art of Not Being Governed*, 210, 327.

39. Vattel, *Law of Nations*, 128–30; Lattimore, *Inner Asian Frontiers*, 130, 238; Sahlins, *Tribesmen*, 1, 3, 38, 81; Scott, *Art of Not Being Governed*.

40. Maine, *Ancient Law*, 91–92.

41. Lattimore, *Mongols of Manchuria*, 76.

42. Lattimore, 76.

43. Lattimore, *Manchuria: Cradle of Conflict*, 48.

44. See, for example, Boris Vladimirtsov, *Obshestvennyi stroi Mongolov: Mongol'skii Kochevoi Feodalizm*. (Izdatel'stvo Akademii Nauk SSSR, 1934).

45. Christopher Dawson, *The Mongol Mission: Narratives and Letters of the Franciscan in Mongolia and China in the Thirteenth and Fourteenth Centuries* (Sheed and Ward, 1955); Lattimore, *Mongolia: Cradle of Conflict*, 73–74; Owen Lattimore, "Open Door or Great Wall?," in *Studies in Frontier History: Collected Papers 1928–1958* (1934; Oxford University Press, 1962), 74.

46. Igor de Rachewiltz, *Index to the Secret History of the Mongols*, vol. 121 of Uralic and Altaic Studies (Indiana University Publications, 1972), §§14, 152, 219, 272.

47. Rachewiltz, *Index to the Secret History*, §65.

48. Rachewiltz, §§207, 219, 255, 279.

49. Rachewiltz, §§121, 279.

50. Rachewiltz, §255; Ata-Malik Juvaini, *The History of the World-Conqueror*, trans. John Andrew Boyle (Harvard University Press, 1958), 1:42–43.

51. Mark Elliott, "Frontier Stories: Periphery as Center in Qing History," *Frontiers of History in China* 9, no. 3 (2014): 336–60, quotation on 336.

52. Elliott, "Frontier Stories," 336–37.

53. Lattimore, *Inner Asian Frontiers*, 552.

54. Munkh-Erdene, *Nomadic Leviathan*, 15–22, 164–76; Lattimore, "Open Door or Great Wall?," 78.

55. Lhamsuren Munkh-Erdene, "The 1640 Great Code: An Inner Asian Parallel to the Treaty of Westphalia," *Central Asian*

Survey 29, no. 3 (2010): 269–88; Munkh-Erdene, *Taiji Government*, 183–234, 416–20, 79–121.

56. Munkh-Erdene, "1640 Great Code."

57. For more on the Döchin and Dörben, see Munkh-Erdene, *Taiji Government*, 416–65.

58. Munkh-Erdene, *Taiji Government*, 416.

59. Munkh-Erdene, "1640 Great Code."

60. Harlan J. Bushfield, "The United Nations Charter: Extension of Remarks of Hon. Harlan J. Bushfield of South Dakota in the Senate of the United States." in *Congressional Record: Proceedings and Debates of the 79th Congress First Session*, appendix volume 91, part 12, June 11, 1945, to October 11, 1945 (pages A2767 to A4294) (United States Government Printing Office, 1945), A3328–29, quotation on A3329.

61. Winston Churchill, Commons. Feb. 28, 1906. 152 Parl. Deb. (4th ser.) (1906) col. 1239.

62. Munkh-Erdene, *Taiji Government*, 181–234.

63. Munkh-Erdene, 436–37.

64. Munkh-Erdene, 169–234.

65. Henry Serruys, "The Office of Tayisi in Mongolia in the Fifteenth Century," *Harvard Journal of Asiatic Studies* 37, no. 2 (1977): 357; Hidehiro Okada, "Dayan Khan as a Yuan Emperor: The Political Legitimacy in 15th Century Mongolia," *Bulletin de l'École française d'Extrême-Orient* 81 (1994): 52.

66. Munkh-Erdene, *Taiji Government*, 142, 198.

67. This is analogous to medieval Mongolian "princes who rule *ulus*" (*ulus medekün kö'üt*) and "the princes who do not rule *ulus*" (*ulus ba ülü medekün kö'üt*). Lhamsuren Munkh-Erdene, "Mongol State Formation and Imperial Transformation," in *The Mongol World*, ed. Timothy May and Michael Hope (Routledge, 2022): 351–69. See also Munkh-Erdene, *Taiji Government*, 207–34.

68. Munkh-Erdene, *Taiji Government*, 76–121.

69. Munkh-Erdene, 107.

70. Munkh-Erdene, 325–34.

71. Munkh-Erdene, 328.

72. Munkh-Erdene, 328.

73. Munkh-Erdene, 333.

74. This is precisely what Injannashi, a nineteenth-century Mongolian historian, found abnormal; Mongolian *khoshuu*s were neither regular subjects of the Qing nor external equal states. Munkh-Erdene, *Taiji Government*, 63–64. Injannashi came to the conclusion that each Mongolian princely *khoshuu* was a state in itself, albeit a tributary to the Daiching Ulus, while all Mongolian princely *khoshuu*s collectively formed the Mongol Ulus, itself a sovereign yet tributary state under the Daiching Ulus. In advancing his claims, Injannashi took what Wheaton categorized as a "tributary or vassal state" as a benchmark. Henry Wheaton, *Elements* (B. Fellowes, 1836), 1:64. Yet Wheaton's definition of sovereign state is essentially that of Vattel; Vattel's own definition qualified Mongolian principalities as sovereign states, even as his "natural law of cultivation" dictated that the Mongols be extirpated. Wheaton, *Elements of International Law*, 62–104; Vattel, *Law of Nations*, 81–85.

75. Munkh-Erdene, *Taiji Government*, 525–28, quotation on 526.

76. Munkh-Erdene, 525.

77. Rachewiltz, *Index to the Secret History*, §52.

78. Rachewiltz, §121.

79. Rachewiltz, §123. See also Igor de Rachewiltz, *The Secret History of the Mongols: A Mongolian Epic Chronicle of the Thirteenth Century* (Brill, 2004), 49.

80. Rachewiltz, *Index to the Secret History*, §§133, 136, 246.

81. Rachewiltz, §§254, 255, 269.

82. Rachewiltz, §§153, 193, 199, 227, 278.

83. Rachewiltz, §§227, 278.

84. Rachewiltz, §202.

85. Rachewiltz, §170.

86. Rachewiltz, §277.

87. Rachewiltz, §199.

88. Rachewiltz, §§186, 224, 242, 243.

89. Rachewiltz, §§255, 270. See also Munkh-Erdene, "Mongol State Formation."

90. Rachewiltz, *Index to the Secret History*, §§252, 260.

91. Weber, *Economy and Society*, 234, 232.

92. Charles R. Beazley, ed., *The Texts and Versions of John de Plano Carpini and William de Rubriquis as Printed for the First Time by Hakluyt in 1598 Together with Some Shorter Pieces* (J.

and C. F. Clay at the University Press, for the Hakluyt Society, 1903), 86–87; see also 58–59.

93. Bodin, *On Sovereignty*, 7–8.

94. Bodin, 8. See also Munkh-Erdene, *Nomadic Leviathan*, 434.

Chapter 3

1. The scholarship on the 1136 map is discussed in Peter Bol, "Exploring the Propositions in Maps: The Case of the 'Yujitu' of 1136," *Journal of Song-Yuan Studies* 46 (2016). I am particularly indebted to Yan Tingting 闫婷婷, "'Yuji tu' yanjiu zongshu 《禹迹图》研究综述—Literature Review of Yu Trace Map," *Hexi xueyuan xuebao* 河西学院学报 32, no. 1 (2016). To view the map in greater detail, go to http://id.lib.harvard.edu/images/olvwork271685/catalog.

2. Akin and Mumford note that "the placement of sites on the east-west axis was substantially less accurate. . . . However[,] distances along the east-west axes, presumably based on terrestrial surveys, are quite accurate." Alexander Akin and David Mumford, "'Yu laid out the lands': Georeferencing the Chinese *Yujitu* [Map of the Tracks of Yu] of 1136," *Cartography and Geographic Information Science* 39, no. 3 (2012): 154–69, quotation on 167. Cheng Yinong has shown that relying on the travel distances between prefectural seats as reported in sources from the time would account for the most obvious discrepancies between the map and the actual geography along the eastern coast, the most accurate part of the map. See Cheng Yinong 成一农, *"Fei ke xue" de Zhongguo chuan tong yu tu* 非科学"的中国传统舆图 (Zhongguo she hui ke xue chu ban she, 2016).

3. Denis Wood and John Fels, *The Power of Maps*, Mappings (Guilford Press, 1992), 51–52.

4. Ouyang Min, 歐陽忞 *Yudi guangji* 輿地廣記, ed. Li Yongxian 李勇先 and Wang Xiaohong 王小红 (Chengdu Sichuan daxue chubanshe 2003).

5. Sima Guang's 司馬光 (1019–1086) chronological history, the *Comprehensive Mirror for Aid in Government*, covers the 1,400 years prior to the Song founding in 960 CE, from the moment in 403 BCE when in Sima's view the ancient Zhou dynasty lost the

way. Shao Yong's 邵雍 (1011–1077) *Book of the Supreme Ultimate for Governing the World* periodizes all history based on a claim about principles of the universe from the beginning of time. The commentaries on the Confucian classics by Wang Anshi 王安石, Su Shi 蘇軾, Cheng Yi 程頤, and others claim to rediscover original meanings that had been ignored since antiquity.

6. Nicolas Tackett, "The Great Wall and Conceptualizations of the Border under the Northern Song," *Journal of Song-Yuan Studies*, no. 38 (2008).

7. Shui Anli, 稅安禮 *Lidai dili zhizhang tu* 歷代地理指掌圖 (Shanghai guji chubanshe, 1989).

8. Tuotuo, ed., *Song shi* 宋史 (Zhonghua shuju, 1977), 475.13793–802.

9. Cao Wanru dates it to 1117 and notes that it is printed upside down, a clear indication that the engraving was for the purpose of reproduction. She provides a transcription of the map's extensive descriptions of foreign states and peoples. See Cao Wanru 曹婉如, "Youguan *Huayi tu* wenti de tantao 有关华夷图问题的探讨," in *Zhongguo gudai ditu ji* 中國古代地圖集, ed. Cao Wanru 曹婉如 (Wenwu chubanshe, 1990). To see the map in detail, go to http://id.lib.harvard.edu/images/olvwork353238/catalog.

10. My original supposition was that the maps were part of the 1596 edition of the gazetteer that was incorporated into the 1640 edition. However, the "West River Bridge" that appears on this map was built only in 1639. The gazetteer states: "In the jimao year [1639] of the Chongzhen reign period[,] the county magistrate Xiong Renlin contributed his own salary to build the West River Bridge here." 崇禎己卯知縣熊人霖捐俸資造西江橋于此 Xiong Renlin 熊人霖, ed., *Yiwu xianzhi* 義烏縣志, 20 vols. (China: s.n., 1640), 3:10a.

11. Xiong Renlin 熊人霖, *Yiwu xianzhi* 義烏縣志, 例解.

12. Xiong Renlin 熊人霖, *Yiwu xianzhi* 義烏縣志, 圖說.

13. There were editions in 1078–85, 1265–74, 1353, 1445, and 1596. Huang Jin's 黃溍 preface to the 1353 edition (included in all later editions) states it is the first with maps.

14. Joseph Dennis pointed me to these rules (纂脩志書凡例 永樂十六年頒降) as found in Wu Zongqi 吳宗器, ed., *Xinxian zhi* 莘縣志.

15. Pan Shutang 潘樹棠 and Li Ruwei 李汝為, eds., *Yongkang xianzhi* 永康縣志, 16 vols. (1892).

16. "Zhejiang quansheng yutu bing shu lu dao li ji 浙江全省輿圖並水路道里記," (1894). The county map of Yiwu can be viewed at https://iiif.lib.harvard.edu/manifests/view/drs:493636145$3i.

17. Some notable works in English include: Maurice Freedman, *Chinese Lineage and Society: Fukien and Kwangtung*, Monographs on Social Anthropology (Athlone Press; University of London, 1966); Patricia Buckley Ebrey and James L. Watson, introduction to *Kinship Organization in Late Imperial China, 1000–1940*, ed. James L. Watson and Patricia Buckley Ebrey (University of California Press, 1986); Robert P. Hymes, "Marriage, Descent Groups, and the Localist Strategy in Sung and Yuan Fu-chou," in *Kinship Organization in Late Imperial China, 1000–1940*, ed. Patricia Buckley Ebrey and James L. Watson (University of California Press, 1986); Michael Szonyi, *Practicing Kinship: Lineage and Descent in Late Imperial China* (Stanford University Press, 2002); David Faure, *Emperor and Ancestor: State and Lineage in South China* (Stanford University Press, 2007); and Xi He, *Lineage and Community in China, 1100–1500: Genealogical Innovation in Jiangxi* (Routledge, 2020).

18. See appendix 6 in Peter K. Bol, *Localizing Learning: The Literati Enterprise in Wuzhou, 1100–1600*, Harvard–Yenching Institute Monograph Series 130 (Harvard University Asia Center, 2022).

19. See the twenty-eighth generation of the Lu family in neighboring Dongyang in Yung-chang Tung, "Table 1: Male Population of the Lu Family, by Generation," China Local, n.d., last accessed June 13, 2024, https://chinalocal.omeka.fas.harvard.edu/items/show/1285?collection=6&page=1.

20. Bol, *Localizing Learning*, 204–28.

21. Fang Xiaoru, 方孝孺 *Xunzhi zhai ji* 遜志齋集 (Ningbo chubanshe, 1996), 13.414 宋氏世譜序. Discussed in Bol, *Localizing Learning*, 225–28.

22. F. W. Mote, "Fang Hsiao-ju," in L. Carrington Goodrich and Chao-ying Fang, eds., *Dictionary of Ming Biography, 1368–1644* (Columbia University Press, 1976), 1:426–33. Song Lian

had been rehabilitated in 1514 with the granting of a posthumous title.

23. For a discussion of how the two models have figured in China's modern history, see John E. Schrecker, preface to *The Chinese Revolution in Historical Perspective*, 2nd ed. (Praeger, 2004), xiii–xvii.

24. Tongqin cunzhi bianzuan weiyuanhui, 桐琴村志編纂委員會 *Tongqin Jinshi zongpu* 桐琴金氏宗譜, 2 vols. (Wuyi xian: Zhejiang dazi jichang, 1995), preface.

25. *Guojia* is a neologism from nineteenth-century Japan, where it was used to translate the Western term *state*. It draws on usage in the first millennium CE, when the capital of a hereditary feudal lord was called a *guo* and the seat of the family of their hereditary senior minister was called a *jia*.

Chapter 4

1. Franck Petiteville and Delphine Placidi-Frot, "Multilateral Diplomacy," in *Global Diplomacy: An Introduction to Theory and Practice*, ed. T. Balzacq and W. Snow (Palgrave Macmillan, 2020), 35–47.

2. Daniel O'Quinn, *Engaging the Ottoman Empire: Vexed Mediations, 1690–1815* (University of Pennsylvania Press, 2019), 44–88.

3. For the most recent study of the Ottoman siege of Vienna in 1683, see Kahraman Şakul, *Viyana Kuşatması: Yedi Başlı Ejderin Fendi* (Timaş Yayınları, 2021).

4. For a classical overview of the war between the Ottoman Empire and the Sacra Lega from the Ottoman sources, see İsmail Hakkı Uzunçarşılı, *Osmanlı Tarihi*, vol. III/I (Türk Tarih Kurumu Yayınları, 1951), 444–608; for a literature review in European scholarship, see Richard Franz Kreutel and Abrahamowicz Zygmunt, *Die Türkenkriege in der historischen Forschung* (F. Deuticke, 1983). See also John Stoye, *The Siege of Vienna* (Collins, 1964); Gábor Ágoston, *The Last Muslim Conquest: The Ottoman Empire and Its Wars in Europe* (Princeton University Press, 2021), 460–510; and I. Parvev, *Habsburgs and Ottomans between Vienna and Belgrade, 1683–1739* (Columbia University Press, 1995).

5. James Falkner, *Prince Eugene of Savoy: A Genius for War against Louis XIV and the Ottoman Empire* (Pen & Sword Military, 2022).

6. Defterdar Sarı Mehmed Paşa, *Zübde-ı Vekayiât*, ed. Abdülkadir Özcan (Türk Tarih Kurumu Yayınları, 1995), 621–26; Silâhdâr Fındıklılı Mehmed Ağa, *Nusretnâme*, ed. Mehmet Topal (TÜBA Yayınları, 2018), 383–84, 418–26; Mücteba İlgürel, "Elmas Mehmed Paşa," *TDV İslam Ansiklopedisi* (November 21, 2022), https://islamansiklopedisi.org.tr/elmas-mehmed-pasa; William Mudford, *Memoirs of Prince Eugene, of Savoy* (printed for Ezra Sargent, 1811), 50–51.

7. Virginia Aksan, *Ottoman Wars, 1700–1870: An Empire Besieged* (Pearson & Longman, 2007), 1–28; Caroline Finkel, *Osman's Dream: The History of the Ottoman Empire* (Basic Books, 2005), 293–312. For a critique of the literature of the Ottoman decline, see Cemal Kafadar, "The Question of Ottoman Decline," *Harvard Middle Eastern and Islamic Review* 4 (1997–1998), 30–75.

8. Rifa'at Abou-El-Haj, "The Formal Closure of the Ottoman Frontier in Europe, 1699–1703," *Journal of American Oriental Studies* 89 (1969): 467–75.

9. Karl A. Roider, *Austrian's Eastern Question, 1700–1790* (Princeton University Press, 1982), 1–21; Erica Ianiro, "Venice after Carlowitz: Change and Challenge in Eighteenth-Century Venetian Policy," in *The Treaties of Carlowitz (1699): Antecedents, Course and Consequences*, ed. Colin Heywood and Ivam Parvev (Brill, 2020), 273–77; Dariusz Kołodziejczyk, "Between Universalistic Claims and Reality: Ottoman Frontier in the Early Modern Period," in *The Ottoman World*, ed. C. Woodhead (Routledge, 2012), 205–19; Dariusz Kołodziejczyk, *The Crimean Khanate and Poland–Lithuania: International Diplomacy on the European Periphery (15th–18th Century): A Study of Peace Treaties Followed by Annotated Documents* (Brill, 2011), 190–95.

10. From Colin Heywood and Ivam Parvev, eds., *The Treaties of Carlowitz (1699): Antecedents, Course and Consequences* (Brill, 2020): Colin Heywood, "'This Great Work': Lord Paget and the Processes of English Mediating Diplomacy in the Latter Stages of the Sacra Lega War, 1697–1698," 35–54; Maurits H. van den Boogert, "The Spoils of Peace: What the Dutch Got Out

of Carlowitz," 56–72; and Lother Höbelt, "From Slankamen to Zenta: The Austrian War Effort in the East during the 1690s," 153–75.

11. For a recent compendium on the Karlowitz Congress, see Colin Heywood and Ivam Parvev, eds., *The Treaties of Carlowitz (1699): Antecedents, Course and Consequences* (Brill, 2020).

12. Abou-El-Haj, "Formal Closure of the Ottoman Frontier."

13. Ágoston, *Last Muslim Conquest*, 421.

14. Ali Yaycıoğlu, "Karlofça Ânı: Osmanlı İmparatorluğu 18. Yüzyıla Nasıl Başladı?," *Tarih ve Toplum: Yeni Yaklaşımlar* 18 (2021): 8–55.

15. Suzanne Lalonde, *Determining Boundaries in a Conflicted World* (McGill-Queen's University Press, 2002), 10–23; Marta Lorente Sariñena, "Uti possidetis, ita domini eritis: International Law and the Historiography of the Territory," in *Spatial and Temporal Dimensions for Legal History: Research Experiences and Itineraries*, ed. M. Meccarelli and M. J. Solla Sastre (Max Planck Institute, 2016), 131–71.

16. In Ottoman Turkish, *uti possidetis, ita possidetis* can be translated as *ala halihi zabtında olan zabtında kala/kalsın.*

17. Rifa'at Abou-El-Haj, "Ottoman Attitudes towards Peace Making: The Karlowitz Case," *Der Islam* 51, no. 1 (1974): 131–37; Van den Boogert, "Spoils of Peace," 67–70; Höbelt, "From Slankamen to Zenta" 159.

18. Ágoston, *Last Muslim Conquest*, 413.

19. Jordan Branch, *The Cartographic State: Maps, Territory, and the Origins of Sovereignty* (Cambridge University Press, 2014), 131.

20. Branch, *The Cartographic State*, 120–41.

21. Doreen Massey, *For Space* (Sage, 2005), 106–11.

22. J. Stoye, *Marsigli's Europe, 1680–1730: The Life and Times of Luigi Ferdinando Marsigli, Soldier and Virtuoso* (Yale University Press, 1994), 168.

23. For Marsili's biography, see Stoye, *Marsigli's Europe, 1680–1730*; Mónika F. Molnár, "An Italian Information Agent in the Hungarian Theatre of War: Luigi Ferdiando Marsigli between Vienna and Constantinople," in *Diplomacy, Information Flow and Cultural Exchange*, ed. S. Brzezinski and Áron Zarnóczki (Cambridge Scholar Publishing, 2014), 86–105; and

Güner Doğan, *Sınırlar Üzerinde bir Hayat: İtalyan General Kont Luigi Ferdinando Marsigli ve Osmanlı İmparatorluğu (1679–1732)* (İtalyan Dostluk Derneği, 2020). For Marsili's autobiography, see *Autobiografia di Luigi Ferdinando Marsigli messa in luce nel secondo centenarto dela morte di lui dal Comitato Marsiliano*, ed. E. Lovarini (Nicola Zanichelli, 1930).

24. Rosita D'Amora, "Luigi Fardiando Marsili, Hezārfenn and the Coffee: Texts, Documents and Translations," *Oriente Moderno* 100, no. 1 (2020): 106–19.

25. After the congress in January 1699, Marsili became the Habsburg representative on the multilateral committee for the delimitation between two Ottoman and Habsburg Empires and Venice. During this two-year committee work, Marsili continued his earlier topographic project and completed one of the most detailed borderlands maps in Europe. His massive book on the Ottoman military, *Stato militare dell'Impero Ottomano*, was published posthumously in 1732. Carlotta De Sanctis, "Stato Militare dell'Imperio Ottomanno, Written by Raffaella Gherardi," *Oriente Moderno* 94, no. 1 (2014): 267–68.

26. Stoye, *Marsigli's Europe, 1680–1730*, 168–74; Mónika F. Molnár, "Italian Information Agent," 100–102; Mónika F. Molnár, "Luigi Ferdinando Marsigli e gli Ottomani: La frontiera asburgico-ottomana dopo la pace di Carlowitz," in *La Politica, la Scienza, le Armi. Luigi Ferdinando Marsili e la Costruzione della Frontiera dell'Impero e dell'Europa*, ed. R. Gherardi (CLEUB, 2010), 147–72, esp. 156–58; Güner Doğan, "The Rivers of Ottoman in the Balkans on the reports of Luigi Ferdinando Marsili (1699–1719)," in *Boundary Letters: Ottoman Officials to Luigi Ferdinando Marsili (1699–1719)*, by Mehmet Demiryürek and Güner Doğan (Ankara Birleşik Dağıtım, 2015), 26.

27. The maps and reports prepared by Marsili for the congress are in the Marsili Archives housed at the University of Bologna. Biblioteca Universitaria di Bologna, Fondo Marsili (hereafter, BUB FM), MS 21 ("Ichnographia fortalitiorum limitaneorum . . .") and MS 58 ("Diversi progetti di pace fra li due imperi Cesareo ed Ottomano . . .").

28. "A Summary, or a short survey, of all the difficulties that must be heeded to establish feasible and secure borders between the two empires, Caesarean and Ottoman; the summary of which is divided into four parts, along with their descriptions and maps." BUB FM, MS 58, fol. 141a.

29. BUB FM, MS 58, fol. 146a.

30. BUB FM, MS 58, pp. 72–73.

31. BUB FM, MS 21.

32. Stoye, *Marsigli's Europe, 1680–1730*, 169.

33. Recep Ahıskalı, "Râmi Mehmed Paşa," *TDV İslam Ansiklopedisi*, https://islamansiklopedisi.org.tr/rami-mehmed-pasa (21 November 2022); idem. *Osmanlı Devlet Teşkilatında Reissülkütâblık (XVIII.Yüzyıl)* (Istanbul: TATAV, 2001), 39, 48, 315, 225; Güner Doğan, *Venediklü ile Dahi Sulh Oluna: 17. ve 18. Yüzyıllarda Osmanlı-Venedik İlişkileri* (İletişim, 2017), 111. For the office of the reis efendi, see Virginia Aksan, *The Ottoman Statesman in War and Peace: Ahmed Rasmi Efendi, 1700–1783* (Brill, 1995).

34. There are two known manuscript copies of *Sulhname*: one at Millet Library in Istanbul (Reşid Efendi, nr. 685, hereafter *Sulhname*); and one at Istanbul University, Yazma Eserler (TY, nr. 268). While using the Millet Library copy, I also consulted with Derya Deniz, "Reisülküttab Ramî Mehmed Efendinin Sulhnamesi (Vekâyi'-i Musâlaha) tahlil ve metin" (master's thesis, Bahçeşehir University, 2016); Defterdar Sarı Mehmed Paşa, *Zübde-i Vekayiât*, 645; and Abou-El-Haj, "Ottoman Diplomacy at Karlowitz," *Journal of the American Oriental Society* 87, no. 4 (1967): 154.

35. *Sulhname*, fol. 13a–14b.

36. *Sulhname*, fol. 25a.

37. *Sulhname*, fol. 13a.

38. *Sulhname*, fol. 58a.

39. *Sulhname*, fol. 61b.

40. *Sulhname*, fol. 58a. See also *Karlofça ile Biten Barışın Özeti*, Süleymaniye Kütüphanesi, Esad Efendi 2241, 55–56. This text was analyzed in Doğan, *Venediklü ile Dahi*, 120–22.

41. For Alexander Mavrocordato (1641–1709), see Nestor Camariano, *Alexandre Mavrocordato, le grand drogman: Son activité diplomatique, 1673–1709* (Institute for Balkan Studies, 1970); Feridun Nazif Uzluk, "Diplomat bir tabip (Aleksandros Mavrokordatos)," *Ankara Üniversitesi Tıp Fakültesi: Mecmuası* 19, no. 4 (1966): 1013–25.

42. Haus-, Hof- und Staatsarchiv (Hereafter HHStA), Vienna, Turcica, 170, *Protocollum*, November 13, pp. 17–18 (chapter author's translation from Italian; emphasis by the chapter author).

43. *Sulhname*, fol. 57a.

44. Mübahat Kütükoğlu, *Osmanlı Belgelerinin Dili* (Kubbealtı Akademisi, 1994), 168.

45. *Sulhname*, fol. 58b.

46. "Vaka-yı Maslaha," İstanbul University, Yazma Eserler (manuscripts collection), no. TY, nr. 268, fol. 74b.

47. Ahmet Karamustafa, "Military, Administrative, and Scholarly Maps and Plans," in *Cartography in the Traditional Islamic and South Asian Societies*, vol. 2, book 1 of *The History of Cartography*, ed. J. B. Harley and David Woodward (University of Chicago Press, 1992), 209–27.

48. BUB FM, MS 58. See also Karamustafa, "Military, Administrative, and Scholarly Maps," 215.

49. Katib Celebi, *An Ottoman Cosmography: Translation of Cihannüma*, ed. Gottfried Hagen and Robert Dankoff, trans. Ferenc Csirkés, John Curry, and Gary Leiser (Brill, 2022), 81.

50. *Sulhname*, fol. 58b.

51. *Sulhname*, fol. 59b.

52. *Sulhname*, fol. 58b (chapter author's translation from Ottoman-Turkish).

53. *Sulhname*, fol. 60a.

54. HHStA, Turcica, 170, *Protocollum*, November 13, pp. 17–18 (chapter author's translation form Italian).

55. Stoye, *Marsigli's Europe, 1680–1730*, 175.

56. Abou-El-Haj, "Ottoman Diplomacy at Karlowitz," 503.

57. G. Dávid, "The Eyâlet of Temesvár in the Eighteenth Century," *Oriente Moderno* 18, no. 1 (1999): 113–28; Ömer Gezer, *Kale ve Nefer: Habsburg Serhaddinde Osmanlı Askeri Gücü (1699–1715)* (Kitap Yayınları, 2020), 137–76.

58. BUB FM, MS 58, fol. 141b (chapter author's translation from Latin).

59. *Sulhname*, fol. 65b.

60. *Sulhname*, fol. 66b.

61. *Sulhname*, fol. 67b (chapter author's translation from Ottoman-Turkish).

62. "Treaty of Karlowitz between the Ottomans and Habsburgs, Article 2. Treaty of Peace between the Emperor and Turkey signed at Carlowitz, 26 January 1699," *General Collection of Treatys*, https://documentsdedroitinternational.fr/ressources/TdP/1699-01-26-TraitedeCarlowitz.pdf (21 November 2022).

63. Doğan, *Venediklü ile dahi*, 188–216; Mehmet Demiryürek and Güner Doğan, *Boundary Letters: Ottoman Officials to Luigi Ferdinando Marsili (1699–1719)* (Ankara Birleşik Dağıtım, 2015).

64. Branch, *Cartographic State*.

Chapter 5

1. Territory is often limited to land, but in this chapter, references to territorial conflicts and territorial disputes include the maritime arena because contestations over both land and water are about the status of sovereignty over portions of Earth's surface.

2. Paul R. Hensel, "Contentious Issues and World Politics: The Management of Territorial Claims in the Americas, 1816–1992," *International Studies Quarterly* 45, no. 1 (2001): 81–109; John A. Vasquez, *The War Puzzle Revisited* (Cambridge University Press, 2009); Gary Goertz, Paul Francis Diehl, and Alexandru Balas, *The Puzzle of Peace: The Evolution of Peace in the International System* (Oxford University Press, 2016); Andrew P. Owsiak and Sara McLaughlin Mitchell, "Conflict Management in Land, River, and Maritime Claims," *Political Science Research and Methods* 7, no. 1 (2019): 43–61.

3. Alexander B. Murphy, "Historical Justifications for Territorial Claims," *Annals of the Association of American Geographers* 80, no. 4 (1990): 531–48, quotation on 532.

4. Emilia Justyna Powell and Krista E. Wiegand, "Conflict Management of Territorial and Maritime Disputes," in *What Do We Know about War?*, ed. Sara McLaughlin Mitchell and John A. Vasquez (Rowman & Littlefield, 2021), 196.

5. See, for example, Hensel, "Contentious Issues and World Politics"; Paul F. Diehl and Gary Goertz, *Territorial Changes and International Conflict* (Routledge, 2002); and Paul R. Hensel, Sara McLaughlin Mitchell, Thomas E. Sowers, and Clayton L.

Thyne, "Bones of Contention: Comparing Territorial, Maritime, and River Issues," *Journal of Conflict Resolution* 52, no. 1 (2008): 117–43.

6. See, for example, Paul K. Huth and Todd L. Allee, "Domestic Political Accountability and the Escalation and Settlement of International Disputes," *Journal of Conflict Resolution* 46, no. 6 (2002): 754–90.

7. See, for example, Andrew P. Owsiak and Derrick V. Frazier, "The Conflict Management Efforts of Allies in Interstate Disputes," *Foreign Policy Analysis* 10, no. 3 (2014): 243–64.

8. See, for example, Kenneth A. Schultz, "What's in a Claim? De Jure Versus De Facto Borders in Interstate Territorial Disputes," *Journal of Conflict Resolution* 58, no. 6 (2014): 1059–84.

9. See, for example, Krista E. Wiegand, "Peaceful Dispute Resolution by Authoritarian Regimes," *Foreign Policy Analysis* 15, no. 3 (2019): 303–21.

10. See, for example, Paul R. Hensel and Sara McLaughlin Mitchell, "Issue Indivisibility and Territorial Claims," *Geo-Journal* 64, no. 4 (2005): 275–85; Krista E Wiegand, *Enduring Territorial Disputes: Strategies of Bargaining, Coercive Diplomacy, and Settlement* (University of Georgia Press, 2011); Powell and Wiegand, "Conflict Management of Disputes."

11. See, for example, Gary Goertz and Paul F. Diehl, "Treaties and Conflict Management in Enduring Rivalries," *International Negotiation* 7, no. 3 (2002): 379–98; Hensel and Mitchell, "Issue Indivisibility."

12. Harris Mylonas and Maya Tudor, "Nationalism: What We Know and What We Still Need to Know," *Annual Review of Political Science* 24, no. 1 (2021): 109–32, quotation on 111.

13. Alexander B. Murphy, "Territorial Ideology and International Conflict," in *The Geography of War and Peace*, ed. Colin Flint (Oxford University Press, 2005), 280–96.

14. David Kaplan and Kathryn Hannum, *Nationalism* (Routledge, 2024).

15. Mabel Berezin, "Identity, Narratives, and Nationalism," in *Routledge Handbook of Illiberalism*, ed. András Sajó, Renáta Uitz, and Stephen Holmes (Routledge, 2021), chap. 16.

16. International Crisis Group, *Turkey–Greece: From Maritime Brinkmanship to Dialogue* (Europe Report No. 263, 2021).

17. See David Ricks and Paul Magdalino, eds., *Byzantium and the Modern Greek Identity* (Ashgate, 1998); and Halil Inalcik, *The Ottoman Empire: The Classical Age, 1300–1600* (Phoenix Books, 1973).

18. On the millet system, see Karen Barkey and George Gavrilis, "The Ottoman Millet System: Non-territorial Autonomy and Its Contemporary Legacy," *Ethnopolitics* 15, no. 1 (2016): 24–42.

19. See Alexis G. Papadopoulos and Triantafyllos G. Petridis, *Hellenic Statecraft and the Geopolitics of Difference* (Routledge, 2021).

20. Yannis Hamilakis, *The Nation and Its Ruins: Antiquity, Archaeology, and National Imagination in Greece* (Oxford University Press, 2007).

21. See Fauruk Birtek and Thalia Dragonas, *Citizenship and the Nation State in Greece and Turkey* (Routledge, 2005).

22. Omer Bartov and Eric D. Weitz, *Shatterzone of Empires: Coexistence and Violence in the German, Habsburg, Russian, and Ottoman Borderlands* (Indiana University Press. 2013).

23. Raoul Blanchard, "The Exchange of Populations between Greece and Turkey," *Geographical Review* 15, no. 3 (1925): 449–56.

24. Greece would continue on to cycle through several regime changes after this point in the twentieth century, notably abolishing and then reinstating the monarchy and enduring several coups, dictatorships, and a civil war before becoming the contemporary Third Hellenic Republic.

25. The 2015 EU migrant crisis served to exacerbate tensions. See Evgenia Kouniaki, "Weaponizing Refugees at the Land Borders of Evros: Constructing the Other through Fear and Danger," *Border Criminologies* (blog), Faculty of Law Blogs, University of Oxford, June 3, 2021, https://www.law.ox.ac.uk/research-subject-groups/centre-criminology/centreborder-criminologies/blog/2021/06/weaponizing.

26. Volker Prott, *The Politics of Self-Determination: Remaking Territories and National Identities in Europe, 1917–1923* (Oxford University Press, 2016); Nikiforos P. Diamandouros, Thaleia Dragonas, and Çaglar Keyder, *Spatial Conceptions of the Nation: Modernizing Geographies in Greece and Turkey* (I. B. Tauris, 2010).

27. Beth A. Simmons, *Territorial Disputes and Their Resolution: The Case of Ecuador and Peru* (Peaceworks no. 27, United States Institute of Peace, 1999), 10.

28. Alexander B. Murphy, "The Sovereign State System as Political-Territorial Ideal: Historical and Contemporary Considerations," in *State Sovereignty as Social Construct*, ed. Thomas Biersteker and Cynthia Weber (Cambridge University Press, 1996), 81–120.

29. Nelson Manrique, "Perils of Nationalism: The Peru–Ecuador Conflict," NACLA (North American Congress on Latin America), September 25, 2007, https://nacla.org/article/perils-nationalism-peru-ecuador-conflict.

30. Vargas Lalama and Andres Leonardo, "Ecuador as Victim: The Development of the Discourse on the Territorial Dispute with Peru, 1860–1981" (master's thesis, University of Calgary, 2018), https://prism.ucalgary.ca/bitstream/handle/1880/107717/ucalgary_2018_lalamavargas_andresleonardo.pdf.

31. Manrique, "Perils of Nationalism."

32. Anssi Paasi, "Bounded Spaces in a 'Borderless World': Border Studies, Power and the Anatomy of Territory," *Journal of Power* 2, no. 2 (2009): 213–34.

33. See, for example, Sarah A. Radcliffe, "Imaginative Geographies, Postcolonialism, and National Identities: Contemporary Discourses of the Nation in Ecuador," *Ecumene* 3, no. 1 (1996): 23–42.

34. For more on this tension, see Georg Maier, "The Boundary Dispute between Ecuador and Peru," *American Journal of International Law* 63, no. 1 (1969): 28–46; and Ronald Bruce St. John, *The Ecuador-Peru Boundary Dispute: The Road to Settlement* (Boundary and Territory Briefing 3, no. 1, 1999, International Boundaries Research Unit, University of Durham).

35. Isaiah Bowman, "The Ecuador-Peru Boundary Dispute," *Foreign Affairs* 20, no. 4 (1942): 757–61, quotation on 760.

36. Simmons, *Territorial Disputes and Their Resolution*, 10.

37. Marcel Fortuna Biato, "The Ecuador-Peru Peace Process," *Contexto Internacional* 38, no. 2 (2016): 621–41, esp. 630; Manrique, "Perils of Nationalism."

38. See generally Biato, "Ecuador-Peru Peace Process"; St. John, *Ecuador-Peru Boundary Dispute*; and Simmons, *Territorial Disputes and Their Resolution*.

39. Benedict Anderson, *Imagined Communities: Reflections on the Origin and Spread of Nationalism* (Verso, 1983).

40. See, for example, Walker Connor, "Beyond Reason: The Nature of the Ethnonational Bond," *Ethnic and Racial Studies* 16, no. 3 (1993): 373–89; and Marco Antonsich, Michael Skey, Shanti Sumartojo, Peter Merriman, Angharad Closs Stephens, Divya Tolia-Kelly, Helen F. Wilson, and Ben Anderson, "The Spaces and Politics of Affective Nationalism," *Environment and Planning C: Politics and Space* 38, no. 4 (2020): 579–98.

41. See, for example, Seiki Tanaka, "The Microfoundations of Territorial Disputes: Evidence from a Survey Experiment in Japan," *Conflict Management and Peace Science* 33, no. 5 (2016): 516–38.

42. Radcliffe, "Imaginative Geographies"; Sarah A. Radcliffe, "Frontiers and Popular Nationhood: Geographies of Identity in the 1995 Ecuador-Peru Border Dispute," *Political Geography* 17, no. 3 (1998): 273–93.

43. Radcliffe, "Frontiers and Popular Nationhood," 289–90.

44. Radcliffe, "Imaginative Geographies," 30.

45. Franck Billé, "Territorial Phantom Pains (and Other Cartographic Anxieties)," *Environment and Planning D: Society and Space* 32, no. 1 (2014): 163–78.

46. Alexander B. Murphy, "The Modernist Territorial Order and the Ukraine Crisis," *Political Geography* 97 (Virtual Forum: War in Ukraine, 2022), doi.org/10.1016/j.polgeo.2022.102667.

47. Max Fisher, "Word by Word and Between the Lines: A Close Look at Putin's Speech," *New York Times*, February 23, 2022, https://www.nytimes.com/2022/02/23/world/europe/putin-speech-russia-ukraine.html.

48. Margaret Wetherell, *Affect and Emotion: A New Social Science Understanding* (Sage, 2012).

49. Noel Parker and Nick Vaughan-Williams, eds., *Critical Border Studies: Broadening and Deepening the "Lines in the Sand" Agenda* (Routledge, 2013).

50. Radcliffe, "Frontiers and Popular Nationhood," 275.

51. As suggested in Alexander B. Murphy, "International Law and the Sovereign State System: Challenges to the Status Quo," in *Reordering the World: Geopolitical Perspectives on the*

Twenty-First Century, ed. George J. Demko and William B. Wood (Westview Press, 1994), 209–24.

Chapter 6

1. This is certainly the case in my field of political science, evident in review essays—for example, Hendrik Spruyt, "The Origins, Development, and Possible Decline of the Modern State," *Annual Review of Political Science* 5 (June 2002): 127–49; and Tuong Vu, "Studying the State through State Formation," *World Politics* 62, no. 1 (January 2010): 148–75.

2. Unlike the other contributions to this volume, this chapter does not contain any illustrative figures of maps or any other images. This choice reflects the core argument of the chapter: that the state can be understood as a collection of ideas, infrastructures, and representations, with all three on an equal conceptual footing. While map images would illustrate the representational element well, they would underemphasize the interplay of all three.

3. There are enormous bodies of literature—somewhat overlapping—on assemblage theory and actor-network theory. My goal here is to draw on the overall argument about diverse elements combining without relying on the more complex or controversial claims.

4. Jan-Hendrik Passoth and Nicholas J. Rowland, "Actor-Network State: Integrating Actor-Network Theory and State Theory," *International Sociology* 25, no. 6 (November 2010): 818–41, quotation on 818.

5. Passoth and Rowland, "Actor-Network State," 824, discussing Timothy Mitchell, "The Limits of the State: Beyond Statist Approaches and Their Critics," *American Political Science Review* 85, no. 1 (March 1991): 77–96.

6. Patrick Carroll, *Science, Culture, and Modern State Formation* (University of California Press, 2006), 3.

7. Carroll, *Science, Culture, and Modern State Formation*, 17.

8. Carroll, 3, 4.

9. Max Weber, "The Profession and Vocation of Politics," in *Political Writings*, ed. Peter Lassman and Ronald Speirs (1919; Cambridge University Press, 1994), 309–69, quotation on 310–11, italics in the original.

10. The most prominent example of this approach is Quentin Skinner, *The Foundations of Modern Political Thought, Volume One: The Renaissance* (Cambridge University Press, 1978).

11. Julia Costa Lopez, Benjamin de Carvalho, Andrew A. Latham, Ayse Zarakol, Jens Bartelson, and Minda Holm, "Forum: In the Beginning There Was No Word (for It): Terms, Concepts, and Early Sovereignty," *International Studies Review* 20, no. 3 (September 2018): 489–519.

12. Stuart Elden, *The Birth of Territory* (University of Chicago Press, 2013). As noted below, Elden's study also traces territory as a technology, incorporating the material tools of territoriality.

13. Skinner, *Foundations of Modern Political Thought*; Elden, *Birth of Territory*, chap. 7; J. H. Burns, ed., *The Cambridge History of Medieval Political Thought, c.350–c.1450* (Cambridge University Press, 1988).

14. Elden, *Birth of Territory*.

15. As Billé (this volume, chap. 7) points out, "exceptions" like this may be fundamental to the persistence of the ideal-typical notion of sovereignty.

16. Langdon Winner, *Autonomous Technology: Technics-out-of-Control as a Theme in Political Thought* (MIT Press, 1977); Langdon Winner, "Do Artifacts Have Politics?," *Daedalus* 109, no. 1 (Winter 1980): 121–36. Winner's argument has been extensively debated and critiqued within STS; here I simply want to draw on the basic insight, and invert it: if technologies are political, then perhaps *politics is also technological*. For discussion, see Bernward Joerges, "Do Politics Have Artifacts?," *Social Studies of Science* 29, no. 3 (June 1999): 411–31; Steve Woolgar and Geoff Cooper, "Do Artifacts Have Ambivalence? Moses' Bridges, Winner's Bridges and Other Urban Legends in S&TS," *Social Studies of Science* 29, no. 3 (June 1999): 433–49; and Nicholas J. Rowland and Jan-Hendrik Passoth, "Infrastructure and the State in Science and Technology Studies," *Social Studies of Science* 45, no. 1 (February 2015): 137–45.

17. See, among others, Daniel R. McCarthy, *Power, Information Technology, and International Relations Theory: The Power and Politics of US Foreign Policy and the Internet* (Palgrave

Macmillan, 2015); and Jon Lindsay, "Restrained by Design: The Political Economy of Cybersecurity," *Digital Policy, Regulation and Governance* 19, no. 6 (September 2017): 493–514.

18. As in, for example, Thomas Risse and Eric Stollenwerk, "Legitimacy in Areas of Limited Statehood," *Annual Review of Political Science* 21 (May 2018): 403–18, quotation on 404.

19. Peer Schouten, "The Materiality of State Failure: Social Contract Theory, Infrastructure and Governmental Power in Congo," *Millennium: Journal of International Studies* 41, no. 3 (June 2013): 553–74.

20. Michael Mann, "The Autonomous Power of the State: Its Origins, Mechanisms and Results," *European Journal of Sociology* 25, no. 2 (November 1984): 185–213, quotations on 188–89.

21. Michael Mann, "Infrastructural Power Revisited," *Studies in Comparative International Development* 43, nos. 3–4 (December 2008): 355–65, quotation on 358. For more discussion, see Hillel Soifer and Matthias vom Hau, "Unpacking the Strength of the State: The Utility of State Infrastructural Power," *Studies in Comparative International Development* 43, nos. 3–4 (December 2008): 219–30.

22. Chandra Mukerji, "The Territorial State as a Figured World of Power: Strategics, Logistics, and Impersonal Rule," *Sociological Theory* 28, no. 4 (December 2010): 402–24, quotation on 402.

23. Jo Guldi, *Roads to Power: Britain Invents the Infrastructure State* (Harvard University Press, 2012), 3. Guldi leaves ambiguous whether the state *has* infrastructure or *is* itself infrastructural, as noted by Rowland and Passoth, "Infrastructure and the State," 140.

24. See, for example, Bruno Latour, *Science in Action: How to Follow Scientists and Engineers through Society* (Harvard University Press, 1987). Again, I am taking a broad reading of some of the key insights here (particularly the way ANT pushes against the conventional emphasis on ideas), without relying on the entire apparatus of ANT. See also Timothy Mitchell, *Rule of Experts: Egypt, Techno-Politics, Modernity* (University of California Press, 2002); and Passoth and Rowland, "Actor-Network State."

25. Bruno Latour, "Visualization and Cognition: Thinking with Eyes and Hands," *Knowledge and Society* 6, no. 1 (1986): 1–40, quotation on 28.

26. Carroll, *Science, Culture, and Modern State Formation*, 168.

27. Carroll, 143.

28. See, for example, Keller Easterling, *Extrastatecraft: The Power of Infrastructure Space* (Verso, 2014); Paul N. Edwards, "Infrastructure and Modernity: Force, Time, and Social Organization in the History of Sociotechnical Systems," in *Modernity and Technology*, ed. Thomas J. Misa, Philip Brey, and Andrew Feenberg (MIT Press, 2003), 185–225; and Brian Larkin, "The Politics and Poetics of Infrastructure," *Annual Review of Anthropology* 42, no. 1 (October 2013): 327–43.

29. James C. Scott, *Against the Grain: A Deep History of the Earliest States* (Yale University Press, 2017), 144.

30. Michael Mann, *The Sources of Social Power, Volume 1: A History of Power from the Beginning to AD 1760* (Cambridge University Press, 1986); Yale H. Ferguson and Richard W. Mansbach, *Polities: Authority, Identities, and Change* (University of South Carolina Press, 1996).

31. For a broad historical perspective, see Charles S. Maier, *Once within Borders: Territories of Power, Wealth, and Belonging since 1500* (Harvard University Press, 2016).

32. The literature on border infrastructure is vast, particularly in political geography (e.g., David Newman and Anssi Paasi, "Fences and Neighbors in the Postmodern World: Boundary Narratives in Political Geography," *Progress in Human Geography* 22, no. 2 (April 1998): 186–207). For political science examples, see Beth A. Simmons and Michael R. Kenwick, "Border Orientation in a Globalizing World," *American Journal of Political Science* 66, no.4. (October 2022): 853–70; and Ron E. Hassner and Jason Wittenberg, "Barriers to Entry: Who Builds Fortified Boundaries and Why?," *International Security* 40, no. 1 (Summer 2015): 157–90.

33. Mark Shirk, "The Universal Eye: Anarchist 'Propaganda of the Deed' and Development of the Modern Surveillance State," *International Studies Quarterly* 63, nos. 1–2 (June 2019): 334–45.

34. John Agnew, "Borders on the Mind: Re-framing Border Thinking," *Ethics & Global Politics* 1, no. 4 (2008): 175–91, esp. 184.

35. STS research has examined various "representational devices," from material artifacts to models, discourses, and other means. See, for example, Michael E. Lynch and Steve Woolgar, eds., *Representation in Scientific Practice* (MIT Press, 1990); and Catelijne Coopmans, Janet Vertesi, Michael Lynch, and Steve Woolgar, eds., *Representation in Scientific Practice Revisited* (MIT Press, 2014).

36. Jordan Branch, *The Cartographic State: Maps, Territory, and the Origins of Sovereignty* (Cambridge University Press, 2014). This built on work by historians and others on the connection between mapping and national identity, boundary creation, and other processes. See Thongchai Winichakul, *Siam Mapped: A History of the Geo-body of a Nation* (University of Hawai'i Press, 1994); and Peter Sahlins, *Boundaries: The Making of France and Spain in the Pyrenees* (University of California Press, 1989).

37. Mitchell, *Rule of Experts*, chap. 3. This draws on the related work on the role of inscriptions: Latour, "Visualization and Cognition."

38. On statistics and censuses, see, among many others, Alain Desrosières, *The Politics of Large Numbers: A History of Statistical Reasoning*, trans. Camille Naish (Harvard University Press, 1998). On the links between mapping and censuses, see Benedict Anderson, *Imagined Communities: Reflections on the Origin and Spread of Nationalism*, rev. ed. (Verso, 1991); and Carroll, *Science, Culture, and Modern State Formation*.

39. Bruce Curtis, *The Politics of Population: State Formation, Statistics, and the Census of Canada, 1840–1875* (University of Toronto Press, 2001), 36.

40. League of Nations: Treaty Series, vol. CLXV, 1936, no. 3802, https://treaties.un.org/doc/Publication/UNTS/LON/Volume%20165/v165.pdf.

41. Reviewed thoroughly in McCarthy, *Power, Information Technology*.

42. Jordan Branch, "Territory as an Institution: Spatial Ideas, Practices and Technologies," *Territory, Politics, Governance* 5, no. 2 (2017): 131–44.

43. In *The Cartographic State*, I focused on demonstrating how mapping served as an important condition of possibility for the emergence and consolidation of the modern state—particularly, its territorial definition. Here I do not want to argue that my earlier argument was necessarily wrong, but that it could be more productive to move those representations inside the concept of the state. Then the questions concern how those representations emerged, how they interacted with the ideational and infrastructural elements of statehood, and how the state *as a whole* came to be consolidated in its "modern" form.

44. See, most prominently, Joseph R. Strayer, *On the Medieval Origins of the Modern State* (Princeton University Press, 1970)—a book that has been widely influential in historical studies in political science. See also Antony Black, *Political Thought in Europe, 1250–1450* (Cambridge University Press, 1992).

45. See, among others, Skinner, *Foundations of Modern Political Thought*.

46. Joseph Bergin, *The Seventeenth Century: Europe, 1598–1715* (Oxford University Press, 2001), 2.

47. See, for example, C. A. Bayly, *The Birth of the Modern World, 1780–1914* (Wiley-Blackwell, 2003).

48. Jeppe Strandsbjerg, "The Cartographic Production of Territorial Space: Mapping and State Formation in Early Modern Denmark," *Geopolitics* 13, no. 2 (2008): 335–58.

49. In the field of international relations, see, for example, Philip G. Cerny, "Globalization and the Changing Logic of Collective Action," *International Organization* 49, no. 4 (Autumn 1995): 595–625.

50. For example, the structure of global networks has increased, rather than undermined, the power of some states, such as the United States' unique position as a network hub and the capabilities this has provided. See Henry Farrell and Abraham L. Newman, "Weaponized Interdependence: How Global Economic Networks Shape State Coercion," *International Security* 44, no. 1 (Summer 2019): 42–79.

51. Carroll, *Science, Culture, and Modern State Formation*, 170.

52. This is explored in detail in Billé, this volume, chap. 7.

53. See, for example, the Cyberthreat Real-Time Map at https://cybermap.kaspersky.com/.

54. Henk van Houtum and Rodrigo Bueno Lacy, "The Migration Map Trap: On the Invasion Arrows in the Cartography of Migration," *Mobilities* 15, no. 2 (2020): 196–219.

55. For examples of a number of alternatives (swarms, "the Stack," etc.) that capture powerful aspects of contemporary organization but—so far, at least—have failed to produce evocative representations, see Billé, this volume, chap. 7.

56. Branch, *Cartographic State*, chap. 3.

57. Matthew Zook, "The Geographies of the Internet," *Annual Review of Information Science and Technology* 40, no. 1 (2006): 53–78.

58. Jordan Branch, "What's in a Name? Metaphors and Cybersecurity," *International Organization* 75, no. 1 (Winter 2021): 39–70.

59. Donal McLaughlin, *Origin of the Emblem, and Other Recollections of the 1945 U.N. Conference*, ed. Jennifer Truran Rothwell (1995), available at https://www.cia.gov/static/5fd-68956127416735fce722dcdcbe457/McLaughlinMonograph.pdf.

60. Here I want to follow Ferguson and Mansbach's *Polities*, which focuses on identifying the diverse features of *polities* historically (explicitly avoiding the term *states*).

Chapter 7

1. Ian G. R. Shaw, *Predator Empire: Drone Warfare and Full Spectrum Dominance* (University of Minnesota Press, 2016), 15.

2. Madeleine Reeves, "Signs of Sovereignty: Mapping and Countermapping at an 'Unwritten' Border," in *The Everyday Lives of Sovereignty: Political Imagination Beyond the State* (Cornell University Press, 2021), 217–39, quotation on 236.

3. As was analyzed by Jordan Branch, if the treaties of Westphalia marked some kind of rupture, "the shift to modern uniformly territorial states was not complete until more than a century after 1648." Jordan Branch, *The Cartographic State: Maps, Territory, and the Origins of Sovereignty* (Cambridge University Press, 2014), 30.

4. N. Katherine Hayles, *Unthought: The Power of the Cognitive Unconscious* (University of Chicago Press, 2017), 5.

5. Benjamin H. Bratton, *The Stack: On Software and Sovereignty* (MIT Press, 2015), 3.

6. Bratton, *The Stack*, 4.

7. Lauren Benton writes that the change from portolan charts to modern maps, with the consequent removal of compasses and rhumb lines, led to world maps that portrayed the oceans as a blank expanse. Lauren Benton, *A Search for Sovereignty: Law and Geography in European Empires, 1400–1900* (Cambridge University Press, 2010), 105.

8. See, for example, "2016 Political Ad by Donald J. Trump for President" (video commercial, 2016), available at https://archive.org/details/PolAd_DonaldTrump_5iqfp.

9. Tina Harris, "Lag: Four-Dimensional Bordering in the Himalayas" in *Voluminous States: Sovereignty, Materiality, and the Territorial Imagination*, ed. Franck Billé (Duke University Press, 2020), 78–90.

10. Marco Ferrari, Elisa Pasqual, and Andrea Bagnato, *A Moving Border: Alpine Cartographies of Climate Change* (Columbia University Press, 2018), 21.

11. Branch, *Cartographic State*, 180.

12. Benedict Anderson, *Imagined Communities: Reflections on the Origin and Spread of Nationalism*, rev. ed. (Verso, 1991), 19.

13. Franck Billé and Caroline Humphrey, *On the Edge: Life along the Russia–China Border* (Harvard University Press, 2021), 27–28.

14. Franck Billé, "Doughnut," *Fieldsights*, Theorizing the Contemporary, September 30, 2019, https://culanth.org/fieldsights/doughnut.

15. Bill Keller, "Soviet Aide Admits Maps Were Faked for 50 Years," *New York Times*, September 3, 1988, https://www.nytimes.com/1988/09/03/world/soviet-aide-admits-maps-were-faked-for-50-years.html.

16. Benton, *Search for Sovereignty*, 2.

17. Benton, 3.

18. See Doug Mack, *The Not-Quite States of America: Dispatches from the Territories and Other Far-Flung Outposts of the USA* (W. W. Norton, 2017).

19. Alastair Bonnett, *Off the Map: Lost Spaces, Invisible*

Cities, Forgotten Islands, Feral Places, and What They Tell Us about the World (Aurum, 2014), 219.

20. See Franck Billé, "Jigsaw: Micropartitioning in the Enclaves of Baarle-Hertog/Baarle-Nassau," in *Voluminous States: Sovereignty, Materiality, and the Territorial Imagination*, ed. Franck Billé (Duke University Press, 2020), 217–29.

21. Jason Cons, *Sensitive Space: Fragmented Territory at the India-Bangladesh Border* (University of Washington Press, 2016).

22. On Texas, see Richard V. Francaviglia, *The Shape of Texas: Maps as Metaphors* (Texas A&M University Press, 1995).

23. Catherine Tatiana Dunlop, *Cartophilia: Maps and the Search for Identity in the French–German Borderland* (University of Chicago Press, 2015), 60.

24. A map created by Williwaw Publishing reverses this practice by foregrounding Alaska and including Hawaii and the Lower 48 as insets. See "The United States of America from Alaska's Point of View," Brilliant Maps, March 3, 2023, https://brilliantmaps.com/alaska-usa/.

25. Wayne Chambliss, "Spoofing: The Geophysics of Not Being Governed," in *Voluminous States: Sovereignty, Materiality, and the Territorial Imagination*, ed. Franck Billé (Duke University Press, 2020), 64–77.

26. Jeremy Packer and Joshua Reeves, "Taking People Out: Drones, Media/Weapons, and the Coming Humanectomy," in *Life in the Age of Drone Warfare*, ed. Lisa Parks and Caren Kaplan (Duke University Press, 2017), 261–81, 262.

27. Packer and Reeves, "Taking People Out," 264.

28. Packer and Reeves, 263.

29. See Franck Billé, "Territorial Phantom Pains (and Other Cartographic Anxieties)," *Environment and Planning D: Society and Space* 32, no. 1 (2014): 163–78. See also the extended discussion in Franck Billé, *Somatic States: On Cartography, Geobodies, Bodily Integrity* (Duke University Press, forthcoming).

30. Jussi Parikka, *Insect Media: An Archaeology of Animals and Technology* (University of Minnesota Press, 2010), xi.

31. Parikka, *Insect Media*, 59.

32. Packer and Reeves, "Taking People Out," 275.

33. Stefan Helmreich, *Alien Ocean: Anthropological Voyages in Microbial Seas* (University of California Press, 2009), 16.

34. Shaw, *Predator Empire*, 22.

35. Quoted in Daniel Immerwahr, *How to Hide an Empire: A History of the Greater United States* (Farrar, Straus and Giroux, 2019), 379.

36. Martin Coward, "Networks, Nodes and Deterritorialised Battlespace: The Scopic Regime of Rapid Dominance," in *From Above: War, Violence and Verticality*, ed. Peter Adey, Mark Whitehead, and Alison J. Williams (Oxford University Press, 2013), 95–117, quotation on 111.

37. Matthew Longo, *The Politics of Borders: Sovereignty, Security, and the Citizen after 9/11* (Cambridge University Press, 2018), 223–24; italics in the original.

38. Bratton, *The Stack*, xviii. Bratton defines the Stack as "a transformation in the technical infrastructure of global systems, whereby planetary-scale computation has so thoroughly and fundamentally transformed the logics of political geography in its own image that it has produced new geographies and new territories that can enforce themselves" (375).

39. See Franck Billé, "Auratic Geographies: Buffers, Backyards, Entanglements," *Geopolitics*, 29, no. 3 (2021): 1004–26, https://doi.org/10.1080/14650045.2021.1881490.

40. Timothy Morton, *Hyperobjects: Philosophy and Ecology after the End of the World* (University of Minnesota Press, 2013), 70.

41. Bratton, *The Stack*, 10–11.

42. Bratton, 316.

43. William Rankin, *After the Map: Cartography, Navigation, and the Transformation of Territory in the Twentieth Century* (University of Chicago Press, 2016), 4.

Chapter 8

1. Eduardo Matos Moctezuma, Raúl Barrera, and Lorena Vázquez, "El Huei tzompantli de Tenochtitlan," *Arqueología mexicana* 25, no. 148 (2017): 52–57; Ingrid Trejo Rosas and Lorena Vázquez Vallin, "El Huei Tzompantli de Tenochtitlan," in *Al pie del Templo Mayor de Tenochtitlan: Estudios en honor de Eduardo Matos Moctezuma*, ed. Leonardo López Luján and Ximena Chávez Balderas (El Colegio Nacional, 2019), 2:109–34.

2. A classic study of the phenomenon in the US is Brian W. Dippie, *The Vanishing American: White Attitudes and U.S. Indian Policy* (Wesleyan University Press, 1982). For a more recent study that implicates cartography, see Allan Greer, *Property and Dispossession: Natives, Empires and Land in Early Modern North America* (Cambridge University Press, 2018). For US representations, see Thomas L. Doughton, "Text, Image and the Discourse of Disappearing Indians in Antebellum American Landscape Painting," *Interfaces* 38 (n.d.): 195–222, https://journals.openedition.org/interfaces/323#ftn2.

3. See Benedict Anderson, *Imagined Communities: Reflections on the Origin and Spread of Nationalism* (Verso, 1983).

4. See Johannes Fabian, *Time and the Other: How Anthropology Makes Its Object* (Columbia University Press, 1983).

5. On the oeuvre of García Cubas, see Magali Marie Carrera, *Traveling from New Spain to Mexico: Mapping Practices of Nineteenth-Century Mexico* (Duke University Press, 2011), 144–83.

6. Raymond B. Craib, *Cartographic Mexico: A History of State Fixations and Fugitive Landscapes* (Duke University Press, 2012), 9.

7. On contemporary pilgrimage and world-making among Nahuatl-speaking communities in contemporary Mexico, see Alan R. Sandstrom and Pamela Effrein Sandstrom, *Pilgrimage to Broken Mountain: Nahua Sacred Journeys in Mexico's Huasteca Veracruzana* (University of Colorado Press, 2022).

8. D. Gramling, *The Invention of Monolingualism* (Bloomsbury Academic, 2016), 3, quoted in E. Demuro and L. Gurney, "Languages/Languaging as World-Making: The Ontological Bases of Language," *Language Sciences* 83, no. 101307 (2021): 1–13.

9. Dorothy Tanck de Estrada, *La educacion ilustrada, 1786–1836: Educacion primaria en la ciudad de Mexico* (Colegio de México, 1977); Dorothy Tanck de Estrada, *Pueblos de Indios y Educación En El México Colonial, 1750–1821* (El Colegio de México, 1999); Mary Kay Vaughan, "Primary Education and Literacy in Nineteenth-Century Mexico: Research Trends, 1968–1988," *Latin American Research Review* 25, no. 1 (1990): 31–66.

10. Craib, *Cartographic Mexico*, 46.

11. Rolena Adorno, *The Polemics of Possession in Spanish American Narrative* (Yale University Press, 2007).

12. Jordan Branch, "Mapping the Sovereign State: Technology, Authority, and Systemic Change," *International Organization* 65, no. 1 (Winter 2011): 1–36, quotation on 19.

13. For the reasoning behind this population estimate, see José Luis de Rojas, *Tenochtitlan: Capital of the Aztec Empire* (University Press of Florida, 2012), 50–53.

14. James Lockhart, *The Nahuas after the Conquest: A Social and Cultural History of the Indians of Central Mexico, Sixteenth through Eighteenth Centuries* (Stanford University Press, 1992), 20–28.

15. Barbara E. Mundy, "Mapping the Aztec Capital: The 1524 Nuremberg Map of Tenochtitlan, Its Sources and Meanings," *Imago Mundi* 50 (1998): 1–22.

16. My thanks to Jennifer Nelson for providing me with a Latin translation.

17. Evidence for the emendations comes from the change in the typeface used. On the woodcut map, the mapmaker carved the legends in a heavy Gothic typeface—the same one for the accompanying letter. Gothic fonts were used widely in Nuremberg and in other Northern European printing centers up to this moment. But then the printer added the two texts into the map block in an entirely different typeface. This font is an Italianate (or Roman) one. These Italianate typefaces were just beginning to be adopted among Northern European printers around 1524, so the font was at that moment a visual novelty.

18. J. H. Elliott, "Cortés, Velázquez and Charles V," in *Letters from Mexico*, ed. Anthony Pagden (Yale University Press, 1986), xi–xxxviii.

19. Hernán Cortés, *Cartas y relaciones de Hernan Cortés al emperador Carlos V*, ed. Pascual de Gayangos (A. Chaix, 1866). Translation by Nancy Fitch, published at "Letters from Hernán Cortés," American Historical Association, n.d., last accessed June 15, 2024, https://www.historians.org/teaching-and-learning/teaching-resources-for-historians/teaching-and-learning-in-the-digital-age/the-history-of-the-americas/the-conquest-of-mexico/letters-from-hernan-cortes.

20. A related conquest fiction created somewhat later held

that Moteuczoma mistook Cortés for a returning deity, Quetzalcoatl. On this, see Camilla Townsend, "Burying the White Gods: New Perspectives on the Conquest of Mexico," *American Historical Review* 108, no. 3 (June 2003): 659–87.

21. Mundy, "Mapping the Aztec Capital."

22. We do not know if the Tenochtitlan map was sold as a single sheet, which would augment its public reach; the few known copies have survived because they were bound in with the text, and so protected by the binding. But we do know that the life of this map was greatly extended as it circulated in various iterations for centuries. Thus, the "polemics of possession" that this map represented were able to extend beyond the literati and reach the growing market for cheap printed maps.

23. Ernst Kantorowicz, *The King's Two Bodies: A Study in Medieval Political Theology* (1957; Princeton University Press, 2016).

24. Evidently, only one print of this exists; details are found in Peter Meurer, "Europa Regina: 16th Century Maps of Europe in the Form of a Queen," *Belgeo*, nos. 3–4 (2008): 355–70, http://journals.openedition.org/belgeo/7711.

25. Elke Anna Werner, "Anthropomorphic Maps: On the Aesthetic Form and Political Function of Body Metaphors in the Early Modern Europe Discourse," in *The Anthropomorphic Lens: Anthropomorphism, Microcosmism and Analogy in Early Modern Thought and Visual Arts* (Leiden: Brill, 2015), 251–72, quotation on 270.

26. For bodies on maps, see Werner, "Anthropomorphic Maps"; and Meurer, "Europa Regina."

27. Christian Jacob, *Sovereign Map: Theoretical Approaches in Cartography throughout History*, trans. Tom Conley, ed. Edward H. Dahl (University of Chicago Press, 2006), 160.

28. Valerie Traub, "Mapping the Global Body," in *Early Modern Visual Culture: Representation, Race, and Empire in Renaissance England*, ed. Peter Erickson and Clark Hulse (University of Pennsylvania Press, 2000), 44–97, quotation on 57.

29. Arthur Weststeijn, "Empire in Fragments: Transatlantic News and Print Media in the Iberian World, ca. 1600–40," *Renaissance Quarterly* 74, no. 2 (2021): 528–70; Nina Lamal,

Jamie Cumby, and Helmer J. Helmers, eds., *Print and Power in Early Modern Europe (1500–1800)*, Library of the Written Word 92 (Brill, 2021).

30. Extra-illustrated edition, once owned by M. C. D. Borden, of L. B. Seeley, *Horace Walpole and His World: Select Passages from his Letters* (Seeley, Jackson, and Halliday, 1884), vol. 6, Lewis Walpole Library, Yale University, folio 225 884S, copy 2.

31. The relevant passage is: "On the whole, the most probable view of the origin of the Mexican tribes seems to be the one ordinarily held, that they really came from the Old World, bringing with them several legends, evidently the same as the histories recorded in the book of Genesis. This must have been, however, at a time when they were quite a barbarous, nomadic tribe; and we must regard their civilization as of independent and far later growth." Edward Burnett Tylor, *Anahuac: Or Mexico and the Mexicans, Ancient and Modern*, 2nd ed. (Longmans, Green, Reader and Dyer, n.d. [1877]), 104.

32. On Rodriguez and Codex Rodriguez-Mondragón, see "Sandy Rodriguez: Codex Rodriguez-Mondragón," Riverside Art Museum, n.d., last accessed June 15, 2024, https://riversideartmuseum.org/exhibits/sandy-rodriguez-codex-rodriguez-mondragon/; and "Book 13: After the Conquest—Codex Rodriguez Mondragon," Creative Capital, n.d., last accessed June 15, 2024, https://creative-capital.org/projects/book-13-after-the-conquest-codex-rodriguez-mondragon/. Some of the information comes from a virtual "studio visit" (via Zoom) with the artist in June 2020, as well as conversations between the chapter author and the artist at the Getty Center, Los Angeles, October 4, 2019.

33. On Castillo Deball, see "Mariana Castillo Deball," Guggenheim, n.d., last accessed June 15, 2024, https://www.guggenheim.org/artwork/artist/mariana-castillo-deball; and the artist's own website, https://castillodeball.org/. Much of this information comes from an interview between the chapter author and the artist in April 2021. Forthcoming is a chapter, Barbara E. Mundy, "Cartographic Presence in the Work of Mariana Castillo Deball," in an upcoming catalogue.

34. For cartographies originating in Indigenous and Afro-descendant communities, see Bjørn Sletto, *Radical Cartogra-*

phies: Participatory Mapmaking from Latin America (University of Texas Press, 2020).

Chapter 9

1. Lawrence E. Gelfand, *The Inquiry: American Preparations for Peace, 1917–1919* (Yale University Press, 1976), 141–42.

2. Guntram H. Herb, *Under the Map of Germany: Nationalism and Propaganda, 1918–1945* (Routledge, 1997), 15.

3. "Map: *German Territorial Losses, Treaty of Versailles, 1919*," Holocaust Encyclopedia, United States Holocaust Memorial Museum, n.d., accessed April 21, 2022, https://encyclopedia.ushmm.org/content/en/map/german-territorial-losses-treaty-of-versailles-1919.

4. See Mark Denil, "Cartographic Design: Rhetoric and Persuasion," *Cartographic Perspectives*, no. 45 (2003): 8–67.

5. Anne Godlewska, "The Idea of the Map," in *Ten Geographic Ideas That Changed the World*, ed. Susan Hanson (Rutgers University Press, 1997), 36.

6. Alexander Kent, "Form Follows Feedback: Rethinking Cartographic Communication," *Westminster Papers in Communication and Culture* 13, no. 2 (2018): 96–112, https://doi.org/10.16997/wpcc.296.

7. See Jacques Bertin, *Semiology of Graphics: Diagrams, Networks, Maps* (University of Wisconsin Press, 1983).

8. Jeremy Crampton and John Krygier, "An Introduction to Critical Cartography," *ACME: An International E-journal for Critical Geographies* 4, no. 1 (2006): 11–33.

9. J. B. Harley, "Deconstructing the Map," in *Writing Worlds: Discourse, Text and Metaphor in the Representation of Landscape*, ed. Trevor J. Barnes and James S. Duncan (Routledge, 1992), 231–47.

10. J. B. Harley, "Cartography, Ethics and Social Theory," *Cartographica* 27, no. 2 (1990): 1–23.

11. See Denis Wood, *Rethinking the Power of Maps* (Guilford, 2010); and Jeremy Crampton, "Thinking Philosophically in Cartography: Toward a Critical Politics of Mapping," *Cartographic Perspectives* 41 (Winter 2002): 4–23.

12. See Joe Gerlach, "Mapping as Performance," in *The Routledge Handbook of Mapping and Cartography* (Routledge, 2017), 90–1007; and Rob Kitchin and Martin Dodge, "Rethinking Maps," *Progress in Human Geography* 31, no. 3 (June 1, 2007): 331–44, https://doi.org/10.1177/0309132507077082.

13. Benedict Anderson, *Imagined Communities: Reflections on the Origin and Spread of Nationalism*, rev. ed. (Verso, 1991); Herb, *Under the Map of Germany*, 7.

14. Thongchai Winichakul, *Siam Mapped: A History of the Geo-body of a Nation* (University of Hawai'i Press, 1997), https://uhpress.hawaii.edu/title/siam-mapped-a-history-of-the-geo-body-of-a-nation/.

15. Guntram H. Herb and David H. Kaplan, eds., *Nations and Nationalism: A Global Historical Overview* (ABC-CLIO, 2008), 1:xvii–xx.

16. Morgane Labbé, "Les usages diplomatiques des cartes ethnographiques de l'Europe Centrale et Orientale au XIXe siècle," *Genèses* 3, no. 68 (2007): 25–47.

17. Morgane Labbé, "Les frontières de la nation allemande dans l'espace de la carte, du tableau statistique et de la narration," in *Les espaces de l'Allemagne au XIXe siècle*, ed. Catherine Maurer (Presses universitaires de Strasbourg, 2010), 49–72.

18. Labbé, "Les usages diplomatiques."

19. Michael Heffernan, "The Politics of the Map in the Early Twentieth Century," *Cartography and Geographic Information Science* 29, no. 3 (January 1, 2002): 207–26, https://doi.org/10.1559/152304002782008512.

20. Centre Georges Pompidou, *Cartes et figures de la terre: Exhibition Catalog* (Paris, 1980), 358.

21. Hans-Dietrich Schultz, "Pax Geographica: Räumliche Konzepte für Krieg und Frieden in der geographischen Tradition," *Geographische Zeitschrift* 75 (1987): 1–22; Karl-Georg Faber, "Zur Vorgeschichte der Geopolitik: Staat, Nation und Lebensraum im Denken Deutscher Geographen vor 1914," in *Weltpolitik, Europagedanke, Regionalismus, Festschrift für Heinz Gollwitzer zum 60. Geburtstag*, ed. H. Dollinger, H. Gründer, and A. Hanschmidt (Aschendorff, 1982), 389–406; Gerhard Sandner and Mechtild Rössler, "Geography and Empire in Germany, 1871–1945," in *Geography and Empire*, ed. Anne Godlewska and Neil Smith (Blackwell, 1994), 115–27.

22. See Jakob Spett, *Nationalitätenkarte der östlichen Provinzen des Deutschen Reiches nach den Ergebnissen der amtlichen Volkszählung vom Jahre 1910* (Moritz Perles, 1918).

23. Eugeniusz Romer, *Atlas Polski* (Freytag & Berndt, 1916).

24. Albrecht Penck, "Die Deutschen im Polnischen Korridor," *Zeitschrift der Gesellschaft für Erdkunde zu Berlin*, 1921, 183–84.

25. Albrecht Penck, "Die Polengrenze," *Illustrierte Zeitung*, 1919, 537.

26. Herbert Heyde, "Die Nationalitäten in den deutschen Ostprovinzen: Eine Fälschung schlimmster Art," *Zeitschrift der Gesellschaft für Erdkunde zu Berlin*, 1919, 185; Walter Geisler, "Politik und Sprachenkarten," *Zeitschrift für Geopolitik* 3 (1926): 707.

27. Geisler, "Politik und Sprachenkarten," 705.

28. Penck, "Die Deutschen im Polnischen Korridor," 174.

29. See Paul Langhans, *Alldeutscher Atlas* (Justus Perthes, 1905).

30. The Pan-German Nationalist Karl Christian von Loesch admitted to this in a letter of November 12, 1921, to the German Foreign Office (Politisches Archiv des Auswärtigen Amtes, Bestand Kult VI A, DiA-2, Nr. 11, Bd. 1).

31. See Morgane Labbé, "Eugene Romer's 1916 Atlas of Poland: Creating a New Nation State," *Imago Mundi* 70, no. 1 (2018): 94–113; and Steven Seegel, *Map Men: Transnational Lives and Deaths of Geographers in the Making of East Central Europe* (University of Chicago Press, 2018).

32. Labbé, "Romer's 1916 Atlas of Poland," 101; Seegel, *Map Men*, 168.

33. See Dietrich Schäfer, *Sprachenkarte der deutschen Ostmarken: Übersicht der ortsanwesenden Bevölkerung nach dem Stande am 1. Dezember 1910, entworfen von Dietrich Schäfer* (Verlag von Karl Curtius, 1910).

34. See Erwin Winkler, *Die Siedlungsgebiete der Deutschen in der Tschechoslowakei: unter Mitverwendung der ausschließlich nach amtlichen Quellen auf Grund der Volkszählungsergebnisse vom 1. Dezember 1930 bearbeiteten Nationalitätenkarte von Erwin Winkler: Mit der am 5.10.1938 in Berlin vom Internationalen Ausschuß beschlossenen Besetzungslinie* (Hirzel, 1938).

35. Labbé, "Romer's Atlas of Poland," 99.

36. Penck, "Die Polengrenze."

37. Penck, "Die Deutschen im Polnischen Korridor."

38. R. Milleker, "Über ethnographische Karten als Grundlage geopolitischer Entscheidungen," *Zeitschrift für Geopolitik* 14 (1937): 640.

39. Albrecht Penck, "Deutscher Volks- und Kulturboden," in *Volk unter Völkern, Bücher des Deutschtums* 1, ed. Karl von Loesch and A. H. Ziegfeld (F. Hirt, 1925), 62–73.

40. Wilhelm Volz, *Die völkische Struktur von Oberschlesien* (M. & H. Marcus, 1921), 10.

41. See Hermann Lautensach, "Geopolitik und staatsbürgerliche Bildung," *Zeitschrift für Geopolitik* 1 (1924): 467–76.

42. Albrecht Penck and Hans Fischer, "Der deutsche Volks- und Kulturboden in Europa." Berlin: Verein für das Deutschtum im Ausland, ca. 1925.

43. Guntram H. Herb, "Maps, Power, and Politics," in *The Routledge Handbook of Mapping and Cartography* (Routledge, 2018), 427–38.

44. Herb, "Maps, Power, and Politics."

45. Michael Fahlbusch, Ingo Haar, and Alexander Pinwinkler, *Handbuch der völkischen Wissenschaften: Akteure, Netzwerke, Forschungsprogramme* (De Gruyter Oldenbourg, 2017).

46. Herb, *Under the Map of Germany*, 104–5.

47. Karl Haushofer, "Die suggestive Karte," *Die Grenzboten* 81 (1922): 17–19.

48. Gerhard Groß, *Das Ende des Ersten Weltkriegs und die Dolchstoßlegende: Reclam—Kriege der Moderne* (Reclam Verlag, 2018).

49. Herb, *Under the Map of Germany*.

50. Bruno Latour, "Visualization and Cognition: Thinking with Eyes and Hands," *Knowledge and Society* 6, no. 1 (1986): 1–40.

51. Michael Burleigh, *Germany Turns Eastwards: A Study of "Ostforschung" in the Third Reich* (Cambridge University Press, 1988).

52. On schools being a pillar of nation building, see Ernest Gellner, *Nations and Nationalism* (Cornell University Press, 1983).

53. Klaus Schleicher, "Introduction: Nationalism and

Internationalism: Challenges to Education," in *Nationalism in Education*, ed. Klaus Schleicher (Peter Lang, 1993), 23–24; Gertjan Dijkink, *National Identity and Geopolitical Visions: Maps of Pride and Pain* (Routledge, 1996), 2–3.

54. Seegel, *Map Men*, 230.

55. Bruno Latour, *Reassembling the Social: An Introduction to Actor-Network-Theory* (Oxford University Press, 2005).

56. Heffernan, "Politics of the Map"; Jeremy Crampton, "The Cartographic Calculation of Space: Race Mapping and the Balkans at the Paris Peace Conference, 1919," *Social and Cultural Geography* 7, no. 5 (2006): 731–52.

57. Alexander Murphy, "Territory's Continuing Allure," *Annals of the Association of American Geographers* 103, no. 5 (2013): 1212–26.

58. See, for example, Penck, "Die Deutschen im Polnischen Korridor"; and Penck, "Die Polengrenze."

59. William Rankin, *After the Map: Cartography, Navigation, and the Transformation of Territory in the Twentieth Century* (University of Chicago Press, 2018).

Chapter 10

1. See, for example, Karl Otmar von Aretin, *Das Alte Reich: 1648–1806*, 4 vols. (Klett-Cotta, 1993); Wolfgang Reinhard, *Probleme deutscher Geschichte, 1495–1806: Reichsreform und Reformation, 1495–1555* (Klett-Cotta, 2001); Johannes Burkhardt, *Vollendung und Neuorientierung des frühmodernen Reiches, 1648–1763* (Klett-Cotta, 2006); Barbara Stollberg-Rilinger, *The Emperor's Old Clothes: Constitutional History and the Symbolic Language of the Holy Roman Empire* (Berghahn, 2015); Joachim Whaley, *Germany and the Holy Roman Empire*, 2 vols. (Oxford University Press, 2012); and Peter H. Wilson, *The Holy Roman Empire: A Thousand Years of Europe's History* (Penguin, 2016).

2. Duncan Hardy, *Associative Political Culture in the Holy Roman Empire: Upper Germany, 1346–1521* (Oxford University Press, 2018).

3. Christopher W. Close, *State Formation and Shared Sovereignty: The Holy Roman Empire and the Dutch Republic, 1488–1696* (Cambridge University Press, 2021).

4. Teresa Neumeyer, *Dinkelsbühl: Der ehemalige Landkreis* (Historischer Atlas von Bayern, Franken 1/40, Kommission für bayerische Landesgeschichte, 2018).

5. Falk Bretschneider and Christophe Duhamelle, "Fraktalität: Raumgeschichte und soziales Handeln im Alten Reich," *Zeitschrift für Historische Forschung* 43, no. 4 (2016): 703–46.

6. For an overview, see Riccardo Bavaj, Konrad Lawson, and Bernhard Struck, eds., *Doing Spatial History* (Routledge, 2021); Ian Gregory, Don DeBats, and Don Lafreniere, eds., *The Routledge Companion to Spatial History* (Routledge, 2018); Ian N. Gregory and Alistair Geddes, eds., *Toward Spatial Humanities: Historical GIS and Spatial History* (Indiana University Press, 2014); Richard White, *What Is Spatial History?* (Spatial History Lab Working Paper, 2010); and Anne Kelly Knowles and Amy Hillier, *Placing History: How Maps, Spatial Data, and GIS are Changing Historical Scholarship* (ESRI Press, 2008).

7. For example, by using natural-language processing or computer vision to extract data from large-text corpora or digitized map collections. See Ian Gregory, Christopher Donaldson, Patricia Murrieta-Flores, and Paul Rayson, "Geoparsing, GIS, and Textual Analysis: Current Developments in Spatial Humanities Research," *International Journal of Humanities and Arts Computing* 9, no. 1 (2015): 1–14; and the Machines Reading Maps project, https://machines-reading-maps.github.io/.

8. See, for example, Anne Kelly Knowles, Levi Westerveld, and Laura Strom, "Inductive Visualization: A Humanistic Alternative to GIS," *GeoHumanities* 1, no. 2 (2015): 233–65; and the special section "Qualitative Spatial Representation," *International Journal of Humanities and Arts Computing* 13, nos. 1–2 (2019).

9. See, for example, Yanni Alexander Loukissas, *All Data Are Local: Thinking Critically in a Data-Driven Society* (MIT Press, 2019); and Catherine D'Ignazio and Lauren F. Klein, *Data Feminism* (MIT Press, 2020).

10. See, for example, this list of papers presented at the IEEE Visualization Conference 2021: https://altvis.github.io/.

11. William Rankin, "How the Visual Is Spatial: Contemporary Spatial History, Neo-Marxism, and the Ghost of Braudel," *History and Theory* 59, no. 3 (2020): 311–42, quotation on 338.

12. To reference just a couple of examples: Dietmar Willoweit, *Rechtsgrundlagen der Territorialgewalt* (Böhlau, 1975); and Erwin Riedenauer, ed., *Landeshoheit: Beiträge zur Entstehung, Ausformung und Typologie eines Verfassungselements des Römisch-Deutschen Reiches* (Kommission für Bayerische Landesgeschichte, 1994). For a case study on Franconia, see Neumeyer, *Dinkelsbühl.*

13. Whaley, *Germany and the Holy Roman Empire*, 1:19.

14. See, for example, Wilhelm Janssen, "Landesherrschaft und Territorium—Punkte, Linien und Flächen. Zu den Problemen territorialer Entwicklungskarten," in *Der geschichtliche Atlas der Rheinlande*, ed. Frank Bartsch (Droste, 2010), 33–48; and Neumeyer, *Dinkelsbühl*, 591.

15. Whaley, *Germany and the Holy Roman Empire*, 1:19–20.

16. Bretschneider and Duhamelle, "Fraktalität," 721.

17. See Wolfgang Wüst, "Flickenteppiche als Metapher für Chaos, Föderalismus und Vielfalt: eine landeshistorische Perspektive," *Zeitschrift für Bayerische Landesgeschichte* 83 (2020): 39–60.

18. Wilson, *Holy Roman Empire*, 184.

19. Duncan Hardy, "Were There 'Territories' in the German Lands of the Holy Roman Empire in the Fourteenth to Sixteenth Centuries?," in *Constructing and Representing Territory in Late Medieval and Early Modern Europe*, ed. Mario Damen and Kim Overlaet (Amsterdam University Press, 2022), 29–52, quotation on 49. See, moreover, Jordan Branch, *The Cartographic State: Maps, Territory, and the Origins of Sovereignty* (Cambridge University Press, 2014), 88–91; and Ernst Schubert, *Fürstliche Herrschaft und Territorium im späten Mittelalter* (Oldenbourg, 2006).

20. Branch, *Cartographic State*, 69–99.

21. See Andreas Rutz, *Die Beschreibung des Raums: Territoriale Grenzziehungen im Heiligen Römischen Reich* (Böhlau, 2018).

22. Axelle Chassagnette, "L'Empire en couleurs: Les perceptions de l'espace germanique, de l'Empire et de ses territoires dans les pratiques d'enluminure des cartes à l'époque moderne (Xvie-Xviiie siècle)," in *Le Saint-Empire: Histoire Sociale: (XVIe-XVIIIe Siècle)*, ed. Falk Bretschneider and Christophe Duhamelle (Éditions de la MSH, 2021), 207–21.

23. See Rutz, *Beschreibung des Raums*. For the Swiss Confederation, see Andreas Würgler, "Which Switzerland? Contrasting Conceptions of the Early Modern Swiss Confederation in European Minds and Maps," in *Political Space in Pre-industrial Europe*, ed. Beat Kümin (Ashgate, 2009), 210–12.

24. See Rutz, *Beschreibung des Raums.*

25. Armin Wolf, "Zum Deutschland-Bild in Geschichtsatlanten des 19. Jahrhunderts," in *Geschichtsdeutung auf alten Karten: Archäologie und Geschichte*, ed. Dagmar Unverhau (Harrassowitz, 2003), 255–86. Andreas Rutz pointed to the political motivations that stood behind the strong reliance on fragmentation metaphors among some German regional historians in the 1930s and '40s. See Rutz, *Beschreibung des Raums*, 39–40.

26. Written inventories of a ruler's dominion, for example, often followed a pointillist model that could easily convey competing or combined claims over the same location. See Rutz, *Beschreibung des Raums*, 242.

27. John Agnew, "The Territorial Trap: The Geographical Assumptions of International Relations Theory," *Review of International Political Economy* 1, no. 1 (1994): 53–80. See also Stuart Elden, "Land, Terrain, Territory," *Progress in Human Geography* 34, no. 6 (2010): 799–817.

28. Martin Lewis and Kären Wigen, *The Myth of Continents: A Critique of Metageography* (Berkeley: 1997), 200.

29. Lewis and Wigen, *Continents*, 10.

30. Lewis and Wigen, *Continents*, 11.

31. See, for example, Jeremy Crampton, "GIS and Geographic Governance: Reconstructing the Choropleth Map," *Cartographica* 39, no. 1 (2004): 41–53.

32. Jessica Hullman, "Why Authors Don't Visualize Uncertainty," *IEEE Transactions on Visualization and Computer Graphics* 26, no. 1 (2020): 130–39.

33. See Bretschneider and Duhamelle, "Fraktalität."

34. See, for example, Kommission für geschichtliche Landeskunde in Baden-Württemberg and Landesvermessungsamt Baden-Württemberg, eds., *Historischer Atlas von Baden-Württemberg* (Kommission für Geschichtliche Landeskunde in Baden-Württemberg, 1972–1982); *Historischer Atlas von Bayern*

(Kommission für bayerische Landesgeschichte, 1950–); Edmund Stengel and Uhlhorn, Friedrich, eds., *Geschichtlicher Atlas von Hessen* (Hessisches Landesamt für geschichtliche Landeskunde, 1960–1984); and Willi Alter, ed., *Pfalzatlas* (Pfälzische Gesellschaft zur Förderung der Wissenschaften, 1963–1994).

35. See the map *Hochgerichtsbarkeit um 1792* in Neumeyer, *Dinkelsbühl*.

36. On the vexed relationship of German historiography with spatial and cartographic scholarship, see, for example, Nina Lohmann, "Der 'Raum' in der deutschen Geschichtswissenschaft," *Acta Universitatis Carolinae Studia Territorialia* 3/4 (2010): 47–93.

37. See the Digitale Kartenwerkstatt Altes Reich project, https://digikar.eu/.

38. See the Mapping the Thirty Years War project, https://mappingtyw.web.ox.ac.uk/about.

39. See Luca Scholz, "Deceptive Contiguity: The Polygon in Spatial History," *Cartographica* 54, no. 3 (2019): 206–16.

40. Luca Scholz, *Borders and Freedom of Movement in the Holy Roman Empire* (Oxford University Press, 2020), 108–27.

41. For an example at the scale of individual settlements, see the map of Schmähingen in Gertrud Diepolder and Max Spindler, *Bayerischer Geschichtsatlas* (Bayerischer Schulbuch Verlag, 1969), 105–6.

42. See, for example, Klaus Nippert, *Nachbarschaft der Obrigkeiten: zur Bedeutung frühneuzeitlicher Herrschaftsvielfalt am Beispiel des Hannoverschen Wendlands im 16. und 17. Jahrhundert* (Hahnsche, 2000), 137–72; and Christophe Duhamelle, *La frontière au village: Une identité catholique allemande au temps des Lumières* (Éditions de l'EHESS, 2010). The opposite could also be true: in cases where several authorities insisted on enforcing their claims over the same subjects, the results could be chilling. See, for example, Karl Brunner, *Der pfälzische Wildfangstreit unter Kurfürst Karl Ludwig* (Wagner'sche Universitäts-Buchhandlung, 1896), 53–54.

43. James C. Scott, *The Art of Not Being Governed: An Anarchist History of Upland Southeast Asia* (Yale University Press, 2009), 60.

44. Scott, *The Art of Not Being Governed*, 58–61.

45. See, for example, Benjamin J. Kaplan, *Divided by Faith: Religious Conflict and the Practice of Toleration in Early Modern Europe* (Harvard University Press, 2007), 161–71.

46. See, for example, Lex Heerma van Voss, ed., *Petitions in Social History* (Cambridge University Press, 2002); Hélène Millet, *Suppliques et requêtes: Le gouvernement par la grâce en Occident (XIIe-XVe siècle)* (École française de Rome, 2003); and Andreas Würgler and Cecilia Nubola, eds., *Bittschriften und Gravamina: Politik, Verwaltung und Justiz in Europa* (Duncker & Humblot, 2005).

47. Georg Robert Wiederkehr, *Das freie Geleit und seine Erscheinungsformen in der Eidgenossenschaft des Spätmittelalters* (Juris, 1976), 158–65.

48. Léonard Dauphant, "La rivalité des pardons: Géographie politique de la grâce dans le royaume de France et les Pays-Bas bourguignons, de Charles VI à François Ier," *Revue historique*, no. 665 (2013): 57–88.

49. For a review of the historiography on supplications and petitions in the early modern German lands, see Andreas Würgler, "Bitten und Begehren: Suppliken und Gravamina in der deutschsprachigen Frühneuzeitforschung," in *Bittschriften und Gravamina: Politik, Verwaltung und Justiz in Europa*, ed. Andreas Würgler and Cecilia Nubola (Duncker & Humblot, 2005), 17–52. See also Helmut Neuhaus, "Supplikationen als landesgeschichtliche Quellen: Das Beispiel der Landgrafschaft Hessen im 16. Jahrhundert," *Hessisches Jahrbuch Für Landesgeschichte* 29 (1979): 63–97; Renate Blickle, "Laufen gen Hof. Die Beschwerden der Untertanen und die Entstehung des Hofrats in Bayern. Ein Beitrag zu den Varianten rechtlicher Verfahren im späten Mittelalter und in der frühen Neuzeit," in *Politische Streitkultur in Altbayern: Beiträge zur Geschichte der Grundrechte in der frühen Neuzeit. Politische Streitkultur in Altbayern*, ed. Renate Blickle (De Gruyter Oldenbourg, 2017), 107–32; as well as the other chapters in Würgler and Nubola, *Bittschriften und Gravamina*. On the role of petitioning in a particularly entangled setting, see Nippert, *Nachbarschaft der Obrigkeiten*, 127–37.

50. See Gabriele Haug-Moritz and Sabine Ullmann, eds., *Frühneuzeitliche Supplikationspraxis und monarchische Herrschaft*

in europäischer Perspektive (Österreichische Akademie der Wissenschaften, 2015).

51. See the project "Untertanensuppliken am Reichshofrat Kaiser Rudolfs II. (1576–1612)" led by Gabriele Haug-Moritz and Sabine Ullmann. The larger patterns organizing this record have been described in Thomas Schreiber, "Die Ausübung kaiserlicher Gnadengewalt durch den Reichshofrat: Untertanensuppliken am Reichshofrat Kaiser Rudolfs II (1576–1612)," in *Frühneuzeitliche Supplikationspraxis und monarchische Herrschaft in europäischer Perspektive*, ed. Gabriele Haug-Moritz and Sabine Ullmann (Österreichische Akademie der Wissenschaften, 2015), 215–28.

52. Monica L. Smith, "Networks, Territories, and the Cartography of Ancient States," *Annals of the Association of American Geographers* 95, no. 4 (2005): 832–49, quotations on 835.

53. See Scholz, *Borders and Freedom of Movement*.

54. Christophe Duhamelle, "Drinnen und draußen: Raum und Identität der Exklave im Alten Reich nach dem Westfälischen Frieden," *Trivium* 14 (2013), 5.

55. Duhamelle, "Drinnen und draußen," 6.

56. See, for instance, the case of Palatinate–Neuburg's safe-conduct official in Regensburg in Otto Rieder, *Das pfalzneuburgische Geleite nach Regensburg und in das Kloster Prüfening* (Mayr, 1908), 148–60. Another example is the safe-conduct official of Brandenburg-Ansbach in the free imperial city of Weißenburg. See August Gabler, "Das Brandenburger und Öttinger Geleit im südlichen Franken," *Archiv für Postgeschichte in Bayern* 9 (1957): 123–27, 123–24.

57. Rudolf Fendler, "Geleitstraßen und Postlinien vor der Französischen Revolution," in *Pfalzatlas*, ed. Willi Alter (Pfälzische Gesellschaft zur Förderung der Wissenschaften, 1971), 2:703–32, 712.

58. Luca Scholz, "Protection and the Channelling of Movement on the Margins of the Holy Roman Empire," in *Protection and Empire. A Global History*, ed. Bain Attwood, Lauren Benton, and Adam Clulow (Cambridge University Press, 2017), 13–28.

59. Luca Scholz, "La strada proibita: L'uso delle strade nel Sacro Romano Impero in epoca moderna," *Quaderni Storici* 53, no. 2 (2018): 335–52.

60. Joel Quirk and Darshan Vigneswaran, *Mobility Makes States: Migration and Power in Africa* (University of Pennsylvania Press, 2015), 23.

61. Quirk and Vigneswaran, *Mobility Makes States*, 24.

62. Whaley, *Germany and the Holy Roman Empire*, 1:41.

63. For an example from Eisenach in 1508, see Manfred Straube, *Zum überregionalen und regionalen Warenverkehr im thüringisch-sächsischen Raum, vornehmlich in der ersten Hälfte des 16: Jahrhunderts* (s. l: 1981), 4 vols., 1:38. For an example from the county of Rieneck in 1727, see Bruno Schneider, "Geleitsteine, Kreuze oder Bildstöcke zeigten das Zoll- und Geleitsrecht an," in *1200 Jahre Schaippach mit Zollberg und Hohenroth*, ed. Lotte Bayer (Historischer Verein Gemünden, 2012), 204–26, esp. 210.

64. For a general discussion of the condominium as a form of political organization in early modern Europe, see Alexander Jendorff, "Gemeinsam Herrschen: Das alteuropäische Kondominat und das Herrschaftsverständnis der Moderne," *Zeitschrift für Historische Forschung* 34, no. 2 (2007): 215–42.

65. Heribert Sturm, *Tirschenreuth* (Historischer Atlas von Bayern, Altbayern I/21, Kommission für bayerische Landesgeschichte, 1970).

66. As in the reference map for territorial superiority in Tirschenreuth in Sturm, *Tirschenreuth*, n.p.

67. On the representation of time on maps, see William Rankin, "Mapping Time in the Twentieth (and Twenty-First) Century," in *Time in Maps: From the Age of Discovery to Our Digital Era*, ed. Kären Wigen and Caroline Winterer (University of Chicago Press, 2020), 15–34; and Tiago Luís Gil, "Taking Speed Seriously: Motion, Simultaneity, and Context in Map-Making for Historical Analysis," *Cartography and Geographic Information Science* 48, no. 4 (2021): 320–37.

68. See also Franck Billé, "Auratic Geographies: Buffers, Backyards, Entanglements," *Geopolitics* 29, no. 3 (2021): 1004–26, https://doi.org/10.1080/14650045.2021.1881490.

69. See, for example, Julia Voss, *Darwin's Pictures: Views of Evolutionary Theory, 1837–1874* (Yale University Press, 2010), 61–126.

CONTRIBUTORS

CY ABBOTT (University of Oregon) is a historical geographer focusing on Europe and the Middle East. His research interrogates the active role geographers and other spatial experts in the early twentieth century had in border-delineation processes and in producing cartographic knowledge in service of statecraft. He is pursuing a doctorate in geography at the University of Oregon.

FRANCK BILLÉ (University of California, Berkeley) is an anthropologist and geographer working at the intersection of cartography, sovereignty, and territoriality. His research interest has led to a number of book projects organized around the theoretical foci of the volumetric, the corporeal, and the topological. His recent books include *Voluminous States: Sovereignty, Materiality, and the Territorial Imagination* (2020), *On the Edge: Life along the Russia-China Border* (with Caroline Humphrey, 2021), *[Redacted]: Writing in the Negative Space of the State* (2024), and *Somatic States: On Cartography, Geobodies, Bodily Integrity* (forthcoming).

PETER K. BOL (Harvard University) is a historian of China's cultural elites at the national and local levels from the seventh to the seventeenth century. He is the author of *"This Culture of Ours": Intellectual Transitions in T'ang and Sung China, Neo-Confucianism in History, Localizing Learning,* and *Localizing Learning: The Literati Enterprise in Wuzhou, 1100–1600,* coauthor of *Sung Dynasty Uses of the I-ching,* and coeditor of *Ways with Words* and various journal articles in Chinese, Japanese, and English. He led Harvard's university-wide effort to establish support for geospatial analysis in teaching and research, and in 2005, he was named the first director of the Center for Geographic Analysis.

JORDAN BRANCH (Claremont McKenna College) is a political scientist who works on the origins, features, and consequences of the territorial system of sovereign states and the role of technology and technological change in international politics. His first book, *The Cartographic State* (2014), examines the important but overlooked role played by mapping technologies in the development of modern states. His current research projects focus on information technology and state transformation today, investigating how cybersecurity, drone warfare, and geospatial intelligence and diplomacy are reshaping state institutions.

GUNTRAM H. HERB (Middlebury College) is a geographer whose major publications include *Scaling Identities: Nationalism and Territoriality* (with David Kaplan, 2018), *Cambridge World Atlas* (2009), *Nations and Nationalisms in Global Perspective: An Encyclopedia of Origins, Development, and Contemporary Transitions*, 4 vols. (with David Kaplan, 2008), and *Under the Map of Germany: Nationalism and Propaganda, 1918–1945* (1997). His current digital research project Indigenous Borderlands and Border Rites (www.border-rites.org) seeks to break the silence that surrounds the more than fifty Native nations divided by the US–Canada border.

VALERIE A. KIVELSON (University of Michigan) is a historian of early modern Russia whose research focuses on cartography, empire, visuality, and witchcraft, with interests in politics, gender, and religion in Russia and in comparative perspective. She is the author of several books, including *Cartographies of Tsardom: The Land and Its Meanings in Seventeenth-Century Russia* (2006) and coauthor with Ronald G. Suny of *Russia's Empires* (2016). Most recently, she coedited *Picturing Russian Empire* (2023) with Joan Neuberger and Sergei Kozlov.

BARBARA E. MUNDY (Tulane University) is an art historian whose scholarship dwells on zones of contact between Native peoples and settler colonists as they forged new visual cultures in the Americas. She has been particularly interested in the social construction of space and its imaginary, which was the subject of her first book, *The Mapping of New Spain* (1996).

Her most recent book, *The Death of Aztec Tenochtitlan, the Life of Mexico City* (2015), draws on Indigenous texts and representations to counter a colonialist historiography and to argue for the city's nature as an Indigenous city through the sixteenth century.

LHAMSUREN MUNKH-ERDENE (National University of Mongolia) has published on the theories of the origin and the nature of the state; state and empire building in premodern Eurasia; premodern Eurasian political order and culture; issues of ethnicity; nationalism; and the construction of collective identity in Mongolia and postsocialist transition. He is the author of *The Taiji Government and the Rise of the Warrior State: The Formation of the Qing Imperial Constitution* (2022) and *The Nomadic Leviathan: A Critique of the Sinocentric Paradigm* (2023).

ALEXANDER B. MURPHY (University of Oregon) is a political-cultural geographer. His work focuses on the ways in which the changing political organization of space reflects and shapes ethnic, socioeconomic, and environmental processes; the foundations and consequences of influential geopolitical ideas and assumptions; the role of legal and political arrangements in mediating human-environment relations; and the importance of geographical education for efforts to address critical challenges facing the contemporary world. He is the author of *The Regional Dynamics of Language Differentiation in Belgium* (1988), *Cultural Encounters with the Environment* (edited with Douglas Johnson; 2000), *Geography: Why It Matters* (2018), *Human Geography: People, Place, and Culture*, 12th ed.

(with Erin Fouberg; 2020), and *The European Culture Area*, 7th ed. (with Terry Jordan-Bychkov and Bella Bychkova Jordan; 2020).

LUCA SCHOLZ (University of Manchester) is a scholar interested in analyzing and questioning data in historical inquiry. His work combines archival research, computational methods, and visualization to study spatial history, intellectual history, and environmental history. His first book, *Borders and Freedom of Movement in the Holy Roman Empire* (2020), is a history of free movement in the early modern German lands, one of the most politically fractured landscapes in European history. His current research combines digital and archival evidence in an effort to develop the study of climate and weather in spatial history.

KÄREN WIGEN (Stanford University) teaches Japanese history and the history of cartography. Her first book, *The Making of a Japanese Periphery, 1750–1920* (1995), mapped the economic transformation of southern Nagano Prefecture during the heyday of the silk industry; her second book was *Myth of Continents* (1997), coauthored with Martin Lewis. *A Malleable Map: Geographies of Restoration in Central Japan, 1600–1912* (2010) explored the roles of cartography, chorography, and regionalism in the making of modern Shinano. Her latest book is *Time in Maps: From the Age of Discovery to Our Digital Era*, coedited with Caroline Winterer (Chicago, 2020).

ALI YAYCIOĞLU (Stanford University) is an associate professor of history and the author of *Partners of the Empire: Crisis of the Ottoman Order in the Age of Revolutions* (2016), which reevaluates the Ottoman Empire within the global context of the revolutionary age in the late eighteenth and early nineteenth centuries. He coedits the Stanford Ottoman World Series: Critical Studies in Empire, Nature and Knowledge, and supervises a digital history lab, Mapping the Ottoman Empire, housed in Stanford's Center for Spatial and Textual Analysis (CESTA).

INDEX

Page numbers in italics refer to illustrations.

Abbott, Cy, 8–9
absolutism, 100
actor-network theory (ANT), and modern
state, 94, 96–97, 193n3, 194n24
Afghanistan, 92
Afinogenov, Gregory, 182n16
African Political Systems (*APS*) (Fortes and
Evans-Pritchard), 33–34
Agnew, John, and "territorial trap," 166
Ágoston, Gábor, 63
agriculture, doctrine of superiority of, 32–33
airspace sovereignty, 108
Akin, Alexander, 186n2
Alexander VI (pope), land giveaway to Spanish
and Portuguese monarchs, 125, 145
Altalbenreuth, 176
American Samoa, 112, 114
Amur River: and border between Russian and
Qing Empires, 20–21, 181n14; Remezov's
section of, 22, 23
Anderson, Benedict, 110
Angola, 92
Anna Ioannovna, Empress of Russia, 24, 29
Appleby, John H., 19
artificial intelligence, and territorial manage-
ment and state contours, 115
assemblage theory, and modern state, 94, 193n3
associative political culture, 163
Austria: map of distribution of nationalities in
territory of, 149; "moving border" between
Italy and, 107–8; and Sacra Lega, 63. *See also*

Karlowitz Congress
Awazuhara, Ayami, 140, 142
Aztec (or Mexica) Empire: cartographic erasure
of, 125–28; destruction of by European powers, 121, 125, 127; displacement of through
mapping, 9–10; expression of political sovereignty through architecture of capital city,
121; Florentine Codex, 137, 138–39, 140. *See
also* Moteuczoma; Tenochtitlan

Baarle-Hertog and Baarle-Nassau condo-
minium, 112
Banat valley: under Habsburg control, 74;
between Hungary and Ottoman Transylvania, 72, 73
Beijing Orthodox Mission, 182n16
Beiton, Fedor, map of Enisei River and Kamchatka, 24
Benton, Lauren, 6, 111, 196n7
Berghaus, Heinrich, *Deutschland, Niederlande,
Belgien und Schweiz*, 149
Berlin Conference, 145
Bernhardi, Karl, *Sprachkarte von Deutschland*,
149
Billé, Franck, 166, 178; and exceptions to
sovereignty, 193n15; and inability to imagine
other forms of spatial belonging, 15, 106;
and noncartographic expressions, 9; and
phantom pain from seeing loss of territory
of state, 91; on road networks in contemporary Russia, 98

Bismarck, Otto von, 149
Boden, 176
Bodin, Jean, concept of sovereignty, 44, 184n37
Bohemia, 129
Bol, Peter, 8, 180n23
Bonnett, Alastair, 112
Book of Documents, 47
border conflicts: and critical emphasis on
tangible-geographic, governmental, and institutional circumstances, 80; disputes over
small and uninhabitable pieces of territory,
107; and historically justified territorial
claims, 79; and need for focus on intangible,
symbolic issues, 81; and role of territorial
imaginaries rooted in ethno-nationalist conceptions, 8–9, 79–82, 90–92; Sino–Indian
border conflict, 107. *See also* Ecuador–Peru
territorial dispute; Greco–Turkish territorial
conflict
border exclusion zones, 110
"borderless world," 118
border making: and ethnographic maps, 149;
Hitler–Stalin Line of Partition of Poland, 144,
145; and homogeneously distributed sovereignty, 107–10; inductive method of, 71–76;
infrastructures of, 98; and national self-
determination, 146; and "Nerchinsk border
stone," 22, 182n20; Ottoman approach to at
Karlowitz, 71–76; and political power, 145,
146–49, 155; and *uti possidetis* principle, 62,
64, 68–69, 74–76

borders, "moving," and climate change, 107–8

border walls: and COVID-19 pandemic, 105–6; and laser fences, motion sensors, CCTV, and radars, 107; nonhomogeneity and noncontiguity of, 106; and populism, 105, 106; and strategies to block or filter outsiders, 106; worldwide increase in, 105, 118

border zone, US, *111*, 111–12

boundary-crossing or boundary-erasing threats, 101

Bowman, Isaiah, 88

Branch, Jordan: call for new mode of modeling sovereignty, 9, 15; *The Cartographic State*, 109–10, 195n43, 196n3; and Treaty of Tordesillas, 125

Bratton, Benjamin: and internet as transterritorial society, 117; and Stack, 117, 196n55, 197n38; on state sovereignty over information, 106–7

Braun, Georg, 131–32, 141

Brazil, erosion of territorial rights of Amerindians in Amazon, 136

Bretschneider, Falk, 165, 166, 167

Brown, Wendy, 5

Budjab, Prince, and *The Khoshuu of Gombosüren, the Erdeni Dalai Jasag, the Setsen Khan Aimag*, 43

Butler, Judith, mapping as performative and reiterative, 148

cairns (*oboo*), 40, 41, 42, 43

California, Native settlements in, 1–2, 2–3

Canada: disregard for sovereign rights of First Nations, 136; *Erasing the Line* (Falardeau, Herb, and Talano), 158, *160–61*; segregation of Indigenous peoples in nomenclature and legality, 158

Cao Wanru, 186n9

Carroll, Patrick: definition of state as *plexus*, 94; on modern political world as composed of "engineering states," 97; on postmodern state's "modern" material basis, 101

cartography: and Amerindians, 9–10; binary distinction between objective scientific maps and propaganda maps, 146, 147; cartographic revolution in Europe, 76; and containment, 107–10 (*see also* containment, and territorial sovereignty); and contemporary art, 137, 140–43; and contiguity, 112–14 (*see also* contiguity, and territorial sovereignty); critiques of conventional practice, 8–9; digital historical, 180n22; and figuration, 129; and homogeneity, 110–12, 165 (*see also* homogeneity, and territorial sovereignty); and immobility, 114–16; jigsaw-puzzle convention, 6–7, 166; and Karlowitz Congress, 62; limits of in representing territorial sovereignty, 6, 107–16; Mongolian, 45; and novel representations of historical ambiguities in territoriality, 10–11; and political sovereignty, 121; representation of nonexclusive forms of dominion, 10–11, 163–64; and research on how people read maps, 147; role in colonizing new territories, 107; role in Indigenous dispossession (*see* Indigenous dispossession); role of technology in closing gaps between physical reality and representation, 107–8; scientific, 147; and territorial imaginaries, 91–92; visual force of, 106, 146–47; and visualization for spatial-historical argumentation, 163; and Westphalian thought, 6, 106, 114. *See also* maps; Russian cartography, Petrine era; Russian cartography, post–Petrine era

Castillo Deball, Mariana: Berlin studio, 140, *142*; and costumes worn by participants in *chinelo* dance, 141–42, *141*; and engraving of 1524 map of Tenochtitlan, 140, *141*; *Nuremberg Map of Tenochtitlan*, 137, 140–42, *141*; use of 1524 map of Aztec Tenochtitlan in works, 122–23, 137, 140–41, 143; use of CNC machine, 140; and viewers in space of map, 142; works reflecting relationship between cartography, imagination, and truth claims, 122

Cenepa War, between Ecuador and Peru, 88–89

Chaghadai, 36, 39, 44

Chambliss, Wayne, technologies for mapping subsurface, 114

Charles V, King of Spain and Holy Roman Emperor, 127, 129

Cheng Yinong, 186n2

Chicanos, 142–43

China: Cultural Revolution, 58; debate over centralized or "enfeoffment" systems of government, 57–58; government prohibition of gazetteers and genealogies after 1949, 58, 60; lineages and single-family villages, 55–58, 60; maps defying the official spatial imaginary of state (see *Map of Chongde Township, Yiwu County, Jinhua Prefecture, Zhejiang; Map of the Traces of Yu*); political identity, 180n23; territorial imaginary based on historical-ethnocultural distinctions, 92. *See also* Jurchens' Great Jin dynasty; Manchu Qing dynasty; Ming dynasty; Song dynasty

chinelo dance, 141–42, *141*

Chinggisid sovereignty: *ejen* as ultimate sovereign, 44; linkage of control of people and control of territory, 7, 38; and *nuntuq*, 36; transformed by Dayan Khanids, 38

Chinggis Khan: as *ejen* of *ulus*, 7, 38, 39, 44; naming of Ögödei as next khan, 39–40; and "princes who rule *ulus*," 44, 185n67

chitmahals, 112–13

choropleth maps, 152, 157, 167–68

Civran, Pietro, 65

climate change, anthropogenic, and "moving borders," 107–8

CNC (computer numerical control) machine, 140

Coastal GasLink pipeline (Canada), 136

Colombia, Republic of, emergence from Viceroyalty of New Granada, 87

condominium model of joint dominion, 164, 174, *175*, 176

containment, and territorial sovereignty: "Do You Know the Shape of Japan?," 108–9, *109*;

and laser fences, motion sensors, CCTV, and radars, 107; and "moving border" between Italy and Austria, 107–8; and space beyond the terrestrial, 108–10

contemporary art, and cartography, 137, 140–43

contiguity, and territorial sovereignty, 112–14

continental shelf, and sovereign rights beyond exclusive economic zones, 108

Corsica, 113

Cortés, Hernán: false claim of Moteuczoma's acceptance of Charles V as overlord, 128, 136; landing with Spanish forces on Gulf Coast of Mexico, 126; second letter to Charles V, 127–28; siege of Tenochtitlan, 127

counter-cartography, 9–11; *Codex Rodriguez-Mondragón* (Rodriguez), 137, 142–43; contemporary art and, 10, 137, 140–43; *De los Child Detention Centers, Family Separations, and Other Atrocities* (Rodriguez), 137, 138–39, *140*; *Erasing the Line* (Falardeau, Herb, and Talano), 158, *160–61*; *Nuremberg Map of Tenochtitlan* (Castillo Deball), 137, 140–42, *141*

Coward, Martin, 116

Craib, Raymond, 123, 125

Crampton, Jeremy, 147

critical border studies, 91

critical map studies, 10

cultural anthropology, codification as discipline, 135

Cuzco, map of, 129, *131*

Cvijić, Jovan, 158

cyberspace, 102

Czoernig, Carl Freiherr von, *Ethnographische Karte der Oesterreichischen Monarchie*, 149

Daesh, 102

Daiching Empire: formation of by Inner Asian princes, 36–37, 38, 39; *The Khoshuu of Dugartsembel, the Jorigtu Jasag, the Setsen Khan Aimag, 40*; New Policies, 39

Dari, Grand Duke, and *The Khoshuu of the Bishreltü Jasag, the Setsen Khan Aimag, 41*

data visualization, 163–64

Dauphant, Léonard, spatial distribution of mercy in medieval France, 168–69

Dayan Khan, 38

de Fer, Nicolas, *Les Etats du Czar ou Empereur des Russes en Europe et en Asie*, 22

Delisle, Joseph-Nicolas, *Carte representant la situation et la distance de la Tartarie Orientale jusqaux terres les plus voisines de l'Amerique, dressée pour la conduite des nouvelles navigations que l'on entreprendra sur ces mers*, 28–29, *28*

Dennis, Joseph, 186n14

density, political, and weakened sovereignty, 167–70

Der deutsche Volks- u Kulturboden in Mittel- u Osteuropa, 153, *154*

Derrida, Jacques, 147

Deutscher Klub (Berlin), 155

Deutschland, Niederlande, Belgien und Schweiz (Berghaus), 149

digital historical cartography, 180n22

digital information networks, 102

digital mapping, 158, 163, 202n7

Digital Tokugawa Atlas, 180n22

Döchin and Dörben confederation, 37, *38*

Dodge, Martin, 148

"Do You Know the Shape of Japan?," 108–9, *109*

Dragon Gate Mountain (Chongde township), 54

Drixler, Fabian, 180n22

Duhamelle, Christophe, 165, 166, 167

Dunlop, Catherine Tatiana, *Cartophilia*, 113

East Asian sovereignty, complexity of, 5–6

Ecuador, "imaginative geographies," 89–90; 5-centavo postage stamp, 90, *90*; and Oriente, 89, *90*

Ecuador–Peru territorial dispute: and absence of ethnocultural foundation to territorial claims, 85–86, 89; Brasilia Peace Agreement, 8–9, 85, 88, 89, 90; Cenepa War, 88–89; and nation building through spatial socializa-

tion, 86, 89; in nineteenth and twentieth centuries, 87, 88–89, *88*; and plausible cases of both sides, 87–88; Rio Protocol of 1942, 88, 89; Tiwinza region, *88*, 89. *See also* South America

Efimov, A. V., 182nn24–25

Eger (now Cheb, Czech Republic), 175

Elden, Stuart: and boundary-defined territory of modern states, 95; and territory as technology, 193n12

Electors Palatine, 171, 173

Elliott, Mark, 36

embodied sovereignty, and maps, 128–32; *Evropae Descriptio* (Quad), 129, *130*; *Mexico and Cuzco* (Braun and Hogenberg), 129, 131–32, *131*, 141

empires, pre- and early modern, spatial and legal anomalies, 5–6

enclaves: Baarle-Hertog and Baarle-Nassau, 112–13; *chitmahals*, 112–13; counter- and counter-counter-enclaves, 112–13; medieval, 112

Engels, Friedrich, on social evolution as control over food production, 33

entomological subject, 116–18

Erasing the Line (Falardeau, Herb, and Talano), 158, *160–61*

Erdoğan, President Recip Tayyip, 85, *85*; and map of "Blue Homeland," 85

ethnocultural narratives, territorial imaginaries, and nation building, 8–9, 79–83, 85, 86, 89, 90–92

ethnographic maps: *Ethnographic Map* (Remezov), 16–19, *16*; *Ethnographische Karte der Oesterreichischen Monarchie* (Czoernig), 149

Europe: cartographic revolution, 76; concept of sovereignty, 32–33, 95; Habsburg control of, 129; lack of territorial contiguity in middle ages, 112; modern enclaves, 112

European mapping, and sovereign power, 22, 158; division of Africa during Berlin Conference, 145; division of world between Spain and Portugal, 145; imagined contours of new continents, 145; *Map of Tenochtitlan*

European mapping, and sovereign power (*cont.*)
and the Gulf of Mexico (Cortés), 126, 127–28,
136, 198n17; maps representing politically
fragmented empires as homogeneous, 111;
maps with political titles and claims of
uniform territorial sovereignty, 23, 28–29, 28,
182nn24–25. *See also* Karlowitz Congress;
New Spain maps

Evans-Pritchard, E. E., *African Political Systems*
(*APS*) (with Fortes), 33

evolutionary anthropology: abandonment of
kinship-to-territory transition, 34, 45; and
administrative divisions and bureaucracy as
defining state, 34; view of Mongol political
organization, 35. *See also* Great Divide

exclusive economic zones (EEZ), 101, 108, 112,
113, 117

Extensive Record of the Realm, 48

"extra-illustration," 133–35

Fabian, Johannes, 122

"failed states," 95–96, 101, 103; maps for (see *Map
of Chongde Township, Yiwu County, Jinhua
Prefecture, Zhejiang; Map of the Traces of Yu*)

Falardeau, Vincent, 160–61

Fang Xiaoru, 55–57

Ferguson, Yale H., 196n60

Fichte, Johann Gottlieb, 148

Fischer, Hans, 153

Florentine Codex, 137

Fortes, Meyer, *African Political Systems* (*APS*)
(with Evans-Pritchard), 33

Foucault, Michel, 147

fractals, 166

Frais condominium: between Electorate of
Bavaria and Kingdom of Bohemia, 175–76,
175; *Wechselfrais*, 175–76

France: hexagonal shape, and national space,
113, 118, 148; as logomap, 113–14, 148

France and Germany, resolution of territorial
conflicts, 92

Francis I, 129

free trade corridors, 11, 117

fur, as tribute, 18, 24, 25, 27

Gabon, 81, 92

Ganerbschaft, 174

García Cubas, Antonio, *Carta Etnografica*,
123–25, 124; native-language place names
on, 124–25; zones marking territory of Indigenous groups on, 123–24

Gazetteer of Yiwu County. See *Yiwu County
Gazetteer* of 1640

geobody, 2, 9, 99

German Empire, 149

German *Geopolitik*, 155

German map campaign, during Weimar
Republic: addition of map category "Use of
German in Commerce," 153, 154; charges of
fraud against French and Polish maps, 146;
claiming all territories lost in Treaty of Versailles, 153; effacement of ethnic diversity in
German-speaking lands, 10; efforts to regain
lost territories, 10, 155; falsifications, 153; and
German geographers' influence on national
territory, 158; and German territory severed
from German state, 149–57; and myth of
"the stab in the back" (*Dolchstosslegende*),
155; and social mobilization around idea of
Greater Germany, 156; Stiftung für deutsche
Volks- und Kulturbodenforschung, 155;
using alternative definitions of national
territory, 153

German territory, maps of: *Der deutsche Volks-
u Kulturboden in Mittel- u Osteuropa*, 153,
154; *Die Deutschen im Polnischen Korridor*
(Penck), 152–53, 152, 158, 159; *Geograficzno-
statystyczny Atlas Polski* (Romer), 150–51,
150, 158, 160; *Hitler–Stalin Line of Partition of
Poland*, 144, 145; *Karte des Deutschen Volks-
und Kulturbodens* (Map of German ethnic
and cultural lands) (Ziegfeld), 155–56, 156;
and social mobilization around idea of
Greater Germany, 156

Getty Museum, project to translate Florentine
Codex, 137

Gillis, John, 4

globalization: challenges to nation-states, 100–
101; global trade, and material infrastructure,
103; and increased power of some states,
195n50. *See also* internet

Godlewska, Anne, 147

Godunov, Petr, and map of Tobolsk and all of
Siberia, 19–20, 20

Golygin, Ivan, 24

Google Maps, 114

Gorbitsa river, and Russo–Chinese border,
21–22

Gosel, 176

gravity-measurement technology, 114–15

Great Code of the Mongols, 36–39; adopted by
Great Assembly of Inner Asian princes, 36–
37, 99; discrimination between greater and
lesser principalities, 37–38; and formation
of Döchin and Dörben confederation and
Daiching Empire, 36–37, 38, 39; and guaran-
tee of territorial rights of lesser powers, 7–8;
and idea of independent but collectively
secured principalities, 99; and Lubsang's
invasion of Zasagtu Khanate, 37–38

Great Divide: and people-to-territory transformation of political authority, 31, 45; as product of evolutionary materialist thinking,
31, 45; and Westphalian story, 32. *See also*
territorial sovereignty; tribe sovereignty

Greco–Turkish territorial conflict: based on
claims of succession to Roman Empire,
82; Greco–Turkish War, 83; and Greek
expansion at expense of Ottoman Empire,
83; over land border along Evros (Meriç)
River, 84; and Megali Idea, 83, 83, 84; and
nation-state ideal of Greek state, 82–83; and
nonspatial sovereignty system of Ottoman
millets, 82; and opposing territorial imaginaries, 84; over resource and sovereignty
rights to waters and airspace, 84–85; and
rise of fixed and spatially exclusive national
identities, 82, 83–84; and shift in contested
territory from terrestrial to maritime arenas,

84; and territorial imaginaries grounded in ethnocultural narratives, 8–9, 80, 81–86, 90; Treaty of Lausanne and population exchange, 84; Turkish movement to claim "Blue Homeland" (Mavi Vatan), 84–85, *85*; and 2015 EU migrant crisis, 191n25; and War of Greek Independence, 82

Guam, 114

Guayaquil, 89

Guldi, Jo, 96, 194n23

guojia (state or nation), 58, 187n25

Gürdeniz, Cem, 84

Habsburgs: conquest of Aztec Empire, 125; control of Banat valley, 74; control of Europe, 129; and Peace of Karlowitz, 61–62, 65–67, 74; takeover of Hungary from Ottomans, 63

Halley, Edmond, 19

Handy Maps through the Ages, Map of the Civilized and the Tribes, 49

Hardeck, 176

Harley, J. Brian, on maps as texts to be deconstructed, 147

Haushofer, Karl, 155

Hayles, N. Katherine, 106

Hegel, Georg Wilhelm Friedrich, 97

Henry VIII, 129

Herb, Guntram H., 10, 160–61, 164

Herder, Johann Gottfried von, 148

Hitler–Stalin Line of Partition of Poland, 144, *145*

Hogenberg, Frans, 131–32, 141

Holy Roman Empire: density of petitions in, 169, *170*; directionality and isotropy in maps of, 174; disputed jurisdiction in *territoria inclausa* of Franconia, 167; eighteenth-century maps of, and spatial description of boundaries, 165; Electors Palatine, 171, 173; fractality model for political organization of, 166; Frais condominium shared between Electorate of Bavaria and Kingdom of Bohemia, 175–76, *175*; homogeneous sovereignty challenged by maps of, 174; importance of road infrastructure for territorial control, 97–98, 171, 173; and mapping of nonexclusive forms of dominion, 10–11, 163–67; mercy in, acts of, 168–70, 204n49; network maps for visualizing dominion over roads and rivers, 10–11, 163–67; and patchwork maps representing contiguous, self-contained, and fragmented polities, *162*, 165, 167–68, 171, 176, 203n25, 204n41; places of residence of subjects petitioning Aulic Council, 169, *170*; pointillist dominion in, 165, 167; practice of *Auslaufen* in, 168; regional historical atlas published in Germany, 166–67; representation of political divisions on early modern maps of, 165; and safe conduct, 171, 173, 174; safe-conduct roads between Worms, Würzburg, Strasbourg, and Ulm, 164, *172*, 173–74, *173*; scholarly criticism of mapping as landscape of contiguity and fragmentation, 5, 10, 163, 164–65, 178; "seigneurial" rather than "territorial" logic of imperial states, 171, 173; territorial system, 5; time variance in high jurisdiction over *Wechselfrais* condominium, 176, *177*; visual metaphors of infrastructure and orientation, 171–74, 176; visual metaphors of political density and dilution of authority, 167–70, 176; visual metaphors of time variance, 174–76; visual variables for representing condominia, 164, 174, *175*, 176

Homann, Johann, 22–23

homogeneity, and territorial sovereignty: border making and, 107–10; isotropic territorial sovereignty, 110; lack of in Petrine-era mapping of Siberia, 27–28; "lumpy sovereignty" of Russia, 110; maps challenging notions of, *172*, 173–74, *173*; in United States, and "The 100-mile border zone," 111–12, *111*

Hong Taiji, 39

Huayang Monastery (Chongde township), 54

Hungary, lost by Ottomans to Habsburgs, 63

iasak (tribute), 18, 24, 25, 27

Immerwahr, Daniel, 11

Incas: Cuzco, map of, 129, *131*; destruction of by European powers, 121

India–Bangladesh enclaves (*chitmahals*), 112–13

Indigenous dispossession: cartographers' and anthropologists' portrayal of Indigenous peoples as out of time, 122, 129, 131–36, 141; cartographic erasure of Aztec Empire, 125–28 (see also *Map of Tenochtitlan and the Gulf of Mexico* [Cortés]); disregard for sovereign rights of Indian nations and First Nations by United States and Canada, 121–22, 136; and embodied sovereignty, 128–32; erosion of territorial rights of Amerindians in Amazon, 136; and false claim of Moteuczoma's voluntary abdication in favor of Charles V, 128, 136; and isoline mapping style, 158, *160–61*; and Lattimore's notion of tribe sovereignty in Mongolia, 31, 34–36, 39, 45; and mapping (*see also* Mexico, and Indigenous territoriality); and maps bringing Indigenous sovereignty back into time, 136–43; ~~Ohlone Stanford Lands~~, 1–3, *3*, 7, 122; and "polemics of possession," 125; territorial maps, role of in, 121–22

Indonesia, 92

infrastructure, as component of statehood, 97–98

Inner Asia: colonial reconfiguration of Eurasian steppe as "Inner Asian frontiers of China," 35–36; nomadic sovereignty in, 7, 31; political authority in, 31, 45

internet: and informational surveillance and control by states, 101; as transterritorial society, 117; visual representations of, 102

Iran, territorial imaginary based on historical-ethnocultural distinctions, 92

Islamic State, 102

isoline maps, 151, 157, 158, *160–61*

isotropy, 110, 174

Jacob, Christian, 129

Japan: "Do You Know the Shape of Japan?," 108–9, *109*, 116; territorial imaginary based on historical-ethnocultural distinctions, 92

and state building, 49–51, 58–59. *See also* Liu Yu, Emperor of the State of Great Qi

Map of Yiwu County, Jinhua Prefecture, Zhejiang, 54, 58, *59*, 186n16; country administration and Confucian school at center of, 58, *59*

mapping, refusal to endorse, 8, 68–71, 99

maps: association of sovereignty with territorial map in sixteenth century, 122, 125–32; association with truth and accuracy, 62, 147; automated, 108; cairns (*oboo*), 40; choropleth, 147, 151, 152, 157, 167–68; and claims to territory and nationalism, 128–32, 146–58; and commercial, scientific, artistic, and political goals, 102; of cybersecurity attacks, 101; digital mapping, 158, 163, 202n7; distinction between scientific and propaganda, 147; as emotionally laden symbols, 9; Google Maps, 114; as initially preceding territory, 107, 114; integral role in state formation, 9, 102, 103; isoline, 151, 157, 158, *160–61*; jigsaw-puzzle style of, 6–7, 166; and landscape views, 129; logomaps, 99, 113–14; map art, 137, 140–43; of migration "crises," 101; modification of in line with changes in geography, 107–8; and national identity and boundary creation, 145, 195n36; nationalist mapping, 123–25, 157, 158; before nation-state, 7; network, 171; of new normal, 107; and noncartographic expressions, 9; patchwork, *162, 164, 165,* 167–68, 171, 176; as performative and reiterative, 148; pointillist mapping, 165, 167; and premodern spatial and legal anomalies, 6, 111; as social constructions, 148; as tools of destabilization and critique, 158; as tools of persuasion, 155–56; and *uti possidetis,* 62, 64, 68–69, 76; as visual arguments relying on graphical choices, 163; visual force of, 106, 146–47. *See also* cartography

Marsili, Luigi Ferdinando: borderland maps of Europe, 189n25; map illustrating Ottoman–Habsburg borders according to Peace of Karlowitz, 67, *67*; maps illustrating

Ottoman–Habsburg borderlands and possible boundaries, 61–62, 65–67, 189n27; maps showing plans for fortresses on Ottoman–Habsburg borders, 65, *66, 67*; recommendation that Habsburg authorities establish control of Banat valley and Timişoara, 74

Martonne, Emmanuel de, 158

Masures, 151

Mavrocordato, Alexander, 61; on inductive method for determining borders, 71–72; refusal to start Karlowitz negotiations based on prepared documents, 69–71

Meador, James, 20–21

Mehmed IV, 62

Messerschmidt, D. G., mapping of Siberia, 22–23

Mexico (Borden), 133–35, *134*, 141

Mexico, and Indigenous territoriality: *Carta Etnografica* (García Cubas), 123–25, *124*; and Cortés's map of Tenochtitlan, 127–28, 136, 198n17; and efforts to impose Spanish language on Indigenous communities, 124–25; *Map of Tenochtitlan and the Gulf of Mexico* (Cortés), 122, 125–28, *126*; *Mexico* (Borden), 133–35, *134*, 141; *Mexico and Cuzco* (Braun and Hogenberg), 129, 131–32, *131*, 141; as out of time, 122, 129, 131–36, 141; variations of 1524 map published in Europe, 129–36; *Vetus Mexico* (Ogilby and Montanus), 132–33, *133*

Mexico City (formerly Tenochtitlan): under control of Habsburg monarchs, 131; under control of Moteuczoma, 125; in 1520, *125*; siege of by Cortés, 127

military: changing spatial categories of war, 116; and "full spectrum dominance," 116; and removal of humans from chain of command, 115–16; traditional warfare, 118

millet system, 82

Ming dynasty: conquest by Manchus' Qing dynasty, 51, 57; policies to restore social stability after periods of civil war, 56

Mitchell, Timothy, on mapping in Egypt, 99

Mittelstelle für zwischeneuropäische Fragen

(Leipzig), 155

Molotov, Vyacheslav, 144, 145

Monahan, Erika, 24–25

Mongolia: and coevolution of statecraft practices across Eurasia, 45; and colonial reconfiguration of Eurasian steppe as "Inner Asian frontiers of China," 35–36; and Daiching Empire, 36–37, 38, 39, *40*; declaration of independence, 39; division into princely shares (*qubi*) under Dayan Khanids, 38, 39; fragmentation by division of inheritance and *taiji* government, 39; *khoshuu* established as effective unit of government, 38, 39, *40–43*, 185n74; Lattimore's notion of tribe sovereignty in, 31, 34–36, 39, 45; supposed transition to territorial rule during Daiching era, 34–35; and territorial sovereignty, 7, 38, 39, 44; and what3words mapping system, 108. *See also* Great Code of the Mongols; *taiji* government

Mongol Ulus, 185n74

Montanus, Arnoldus, 132–33

Montevideo Convention, and qualifications for international sovereignty and statehood, 99–100

Morgan, Lewis, 32; on agriculture and civilizations, 33; on ancient and modern political organization, 33, 45

Morton, Timothy, on hyperobjects, 117

Moteuczoma: and acceptance of Charles V as overlord, 128, 136; assassination of, 127; representation of domain on *Map of Tenochtitlan*, 126–27

Mount Similaun (Austria), 108

Mugl, 176

Mukerji, Chandra, 96

Mumford, David, 186n2

Mundy, Barbara, 9–10, 99

Munich Conference, 149, 151

Munkh-Erdene, Lhamsuren, 7–8, 99, 136, 180n23

Murphy, Alexander B., 8–9, 158

Mustafa Kemal (Atatürk), 83–84

Mylonas, Harris, 81

Nahuatl (Aztec language), 123–25, 128, 137, 198n7

nanotechnology, 116

nationalism, and territorial conflict, 8–9, 79–82, 90–92

nationalist mapping: across Europe, 158; and hegemonic "scripts," 157; in nineteenth century, 123–25. *See also* German map campaign, during Weimar Republic

Nationalitäts-Karte von Deutschland (Kiepert), 149

national territory maps, rethinking, 157–58

nation-state, redefining, 93–95, 193n2, 195n43; and alternatives to statehood, 100, 102–3; as discrete and bounded unit, 106; emergence or origins of statehood, 100; and exceptions to statehood, 101–2; and globalization, 100–101; and ideas about organization, authority, and interaction, 95–96; and infrastructure as component of statehood, 96–98; and persistence of statehood, 100; and representations of statehood, 98–100, 101

nation-state building: *Map of the Traces of Yu*, 49–51, 58–59; and schools, 157; and spatial socialization, 86, 89; and territorial imaginaries based on historical-ethnocultural distinctions, 8–9, 79–83, 85, 86, 89, 90–92

nation-state concept: as assemblages and actor-networks, 94, 96–97; and capacity to intervene in society, 95; Carroll's definition of, 94; defined by administrative divisions and bureaucracy, 34; difficulty defining in central and eastern Europe, 148–49; emergence within existing state structure, 148; grounded in historical-ethnocultural territorial imaginary, 8–9, 79–86, 90–92; as imagined community based on belonging to particular group and place, 148; and notion of "failed state," 95–96; and Romantic thought, 148–49; spatial aspect of, 95; Weber's definition of, 4

navigation systems, electronic, 118

Nazi Germany, and Soviet Union, *144*, 145

"Nerchinsk border stone," 22, 182n20

network maps, and ancient states, 171

Neualbenreuth, 176

Neumeyer, Teresa, 167

New Spain maps: *Map of Tenochtitlan and the Gulf of Mexico* (Cortés), 122, 125–28, *126*; *Mexico* (Borden), 133–35, *134, 141*; *Mexico and Cuzco* (Braun and Hogenberg), 129, 131–32, *131, 141*; variations of 1524 map published in Europe, 129–36; *Vetus Mexico* (Ogilby and Montanus), 132–33, *133*

Nigeria, 81

nomadic clans: as having no sovereignty, 7, 31; *uluses* of on maps, 25. *See also* Chinggis Khan; Great Code of the Mongols

nonstate groups controlling territory, 101

nuntuq (territory or country), Chinggisid formation of, 36

Nurhachi, 39

Ogilby, John, 132–33

~~Ohlone~~ *Stanford Lands*, 1–3, *3*, 7, 122

Okinawa, 6

"100-mile border zone, The" 111–12, *111*

Oriente, 89, 90

Ottoman Empire: collapse of, 83; millet system, 82; use of cartography, 70; and War of Greek Independence, 82; war with Sacra Lega, 63

Ottoman mission to Karlowitz Congress: acceptance of European concept of territorial sovereignty, 63–64; border-making process, 71–72; moral-ecological argument refuting *uti possidetis*, 74–76; and popular revolt of 1703, 68; proposal to determine borders with local input, 62; refusal to start negotiations based on prepared documents, 8, 69–71, 99

outer Mongolia (now Mongolian People's Republic), 51

Packer, Jeremy, 115

Paget, William, 61

Parikka, Jussi, 115

Paris Peace Conference, 149

Paris Peace Treaties, 146

patchwork maps: distorting perceptions of political geography, 165; of Holy Roman Empire, 162, 165, 167–68, 171, 176; political density of reframed as dilution of authority, 164

Peace of the Pyrenees, 64

Peace of Westphalia. *See* Westphalia, Peace of

Peace Treaty of Karlowitz, 61, 62

Penck, Albrecht: concept of *Volksund Kulturboden*, 153, 154, 155, 156; criticism of German ethnographic maps, 151; *Die Deutschen im Polnischen Korridor*, 152–53, *152*, 158, *159*; dot maps, 158, *159*

Perdue, Peter, 182n20

persuasion, maps as tools of, 155–56

Peru, lack of nation-building discourse in historical-ethnocultural terms, 86

Peter the Great, 15–16, 19. *See also* Siberia, Petrine maps of

Phoenicians, 171

pixelated subject, 117

pointillist mapping, 165, 167, 203n26

political power, and border mapping, 145, 146–49

Polities (Ferguson and Mansbach), 196n60

populist ethno-nationalism, global rise in, 105, 106, 118

portolan charts, and change to modern maps, 196n7

Potapov, Dmitrii, 24

Powell, Emilia, 80

propaganda maps, 147

Psianchin, A. V., 17

pueblos originarios, 142

Putin, Vladimir, emphasis on ethnocultural continuities between Russians and Ukrainians, 91

Putsch, Johannes: *Europa regina*, 129, 199n24; notion of body of monarch, 129

Qian, Sima, *The Secret History of the Mongols (SHM)*, 35–36, 39, 44

Qin, 47, 48

Quad, Matthias, *Evropae Descripto*, 129, 130

Querenbach, 176

Radcliffe, Sarah, 89–90, 91

Rami Mehmed Efendi, 61; and border making by multilateral commissions in conversation with local people, 71–72; *Peace Treatise* (*Sulhname* or *Vaka-ı Maslaha*), 68, 189n34; refusal to start Karlowitz negotiations based on prepared documents, 70–71; and *uti possidetis*, 75

Rankin, William: *After the Map*, 6–7, 118; and pointillist mapping, 167; and visual strategies for seeing world differently, 164

Ravina, Mark, 6, 180n22

Rechtsstaat's rule of law, 31, 45

Reeves, Joshua, 115

Remezov, Semen Ul'ianovich: *Chertezhnaia kniga* (Sketchbook), 16, 21, 22, 24, 25; copy of Godunov map, 19–20, *20*; *Ethnographic Map*, 16, 16–19; and "free" (*vol'nye*) Tungus clans, 23, 27; granular approach to mapmaking, 18–19, 25–27; and Great Wall, 21; and guard posts as indication of borders, 21; and *iasak* (tribute) status, 18, 24, 25, 27; and Irtysh River, 25, 26; *Khorograficheskaia chertezhnaia kniga* (Chorographic Sketchbook), 16–17, 21–22, 25, 27; and Mangazeia region, 25, 26; maps as counterpoint to conventional representations of territoriality, 7; and section of the Amur, 22, 23; *Sluzhebnaia chertezhnaia kniga* (Working Sketchbook), 17, 19–20; and *uluses* (tents) and *kochev'ias* (camps) of nomadic residents, 25, 27; use of symbols to denote resistance to tsarist control, 24–25

representations, and statehood, 98–100; in legal conventions and concepts, 99; mapping, 98–99, 101; statistics and censuses, 99

Ribbentrop, Joachim von, 144, 145

Rifa'at Abou-El-Haj, 63

Rio Protocol of 1942, 88, 89

Rodriguez, Sandy: assertion of natural world over political order, 142–43; *Codex Rodriguez-Mondragón*, 137, 142–43; *De los Child Detention Centers, Family Separations, and Other Atrocities*, 137, 138–39, *140*; erasure of California–Mexico border, 137, 142; and images from Florentine Codex, 137, 138–39, *140*; representation of dispossession of Mexicans and Mexican Americans, 137–38; works reflecting relationship between cartography, imagination, and truth claims, 122

Roman Empire: concept of empire, applied to kingdoms, 95; display of spoils of war, 121. *See also* Holy Roman Empire

Romer, Eugeniusz, *Geograficzno-statystyczny Atlas Polski*, 150–51, *150*, 158, 160

Russia: border exclusion zones, 110; contemporary violation of Ukrainian sovereignty, 15, 29, 91, 118; "lumpy sovereignty," 110, 165–66; resolution of territorial conflicts with China, 92; territorial imaginary based on historical-ethnocultural distinctions, 92; vertical power reflected in radial roads, 98, 110

Russian and Qing Empires, border between, 20–21, 181n14

Russian cartography, Petrine era: in early eighteenth century, 15, 23; and formal international borders with Poland, 19; and nebulous eastern borders, 19–23; political titles in, absence of, 23; representation of sovereignty over bounded territory in, absence of, 24. *See also* Remezov, Semen Ul'ianovich; Siberia, Petrine maps of

Russian cartography, post–Petrine era, 28–29, *28*

Rutz, Andreas, 165, 203n25

Ruzzini, Carlo, 61, 65

Sacra Lega, victories over Ottomans, 63

safe conduct, in early modernity: linked to roads and rivers, 171; and movement of toll stations with traffic flows, 174; role in channeling merchants, carters, and goods through territories, 173; as tool of territorial control and expansion, 171, 173

Sahlins, Peter, 22

Schachten, 176

Schäfer, Dietrich, language map, 151

Schlick, Count Leopold, 61

Scholz, Luca: and digital mapping, 158; and governance of roads in Holy Roman Empire, 97; and representation of nonexclusive forms of dominion on maps, 10–11; and territorial system of early modern Europe, 5

science and technology studies (STS): applied to study of political institutions, 96, 193n16; and representational devices, 195n35

Scott, James C.: *The Art of Not Being Governed*, 168; role of measurement and writing in "early states," 97; Zomia, 35

Secret History of the Mongols, The, 7

Seegel, Steven, 151, 157

Shao Yong, *Book of the Supreme Ultimate for Governing the World*, 186n5

shared sovereignty, 163. *See also* Holy Roman Empire

Sheehan, James, 5

Sholoi Mahasammata Setsen Khan, 41

Siberia, Petrine maps of: and alternative approach to state and territory, 15–16; ambiguous international borders on, 19; and European cartographers, 22–23; lack of homogeneous and bounded territoriality on, 27–28; and sovereignty, as constructed out of individual relationships, 15–16, 25–27. *See also* Remezov, Semen Ul'ianovich

Sima Guang, *Comprehensive Mirror for Aid in Government*, 186n5

Sino–Indian border conflict, 107

Slezkine, Yuri, 19

Slipchenko, Vladimir, 116

Soja, Edward, "Borders Unbound," 1

Song dynasty: as "central country" surrounded by tribal peoples, 49; and Jurchens, 49; and Mongols, 51

Song Lian, 55, 56

158; segregation of Indigenous peoples in nomenclature and legality, 158, 180n20; territories, and varying integration, representation, and rights, 112

uti possidetis, ita possidetis, 64: and homogeneity of territorial sovereignty, 64; and Karlowitz Congress, 68–69, 74–76; in Ottoman Turkish, 188n16

Utrecht, treaties of, 64

Vattel, Emer de, 185n74

Venice, Republic of, 63

Vetus Mexico (Ogilby and Montanus), 132–33, *133*

violence, and territoriality, 4, 84

Vladimirtsov, Boris, 35

volumetric sovereignty, 108

Volz, Wilhelm, 153, 155

Waldsassen (Bavaria), 175–76

Wang Anshi, 186n5

War of Greek Independence, 82

Weber, Max: on bureaucracy, 34; definition of state, 4, 95, 97; emphasis on territory and state, 183n23; and "sultanism," 44

Weimar Republic. *See* German map campaign, during Weimar Republic

Werner, Elke Anna, 129

West, Jen Ward, ~~Ohlone~~ *Stanford Lands*, 1–3, *3, 7*, 122

Westphalia, Peace of: as conventional birthdate of modern state system, 5, 34; and lands of non-Westphalian rulers, 32, 183n6; and legal equality and noninterference among signatories, 32; as norm and myth, 106, 114; projection of ideas about sovereignty during European colonial era, 32; reframing model of, 117; resultant nation-state territories, 32; use of maps, 64

Whaley, Joachim, 164–65, 174

what3words, 108

Wheaton, Henry, 185n74

Widmer, Eric, 21

Wiegand, Krista, 80

Wigen, Kären, and "contiguity-fetish" of political geography, 166

William of Rubruck, 35

Williwaw Publishing, 197n24

Wilson, Peter: new maps of Thirty Years' War, 167; and notion of fragmentation, 165

Wilson, Woodrow, Fourteen Points, 146

Winichakul, Thongchai. *See* Thongchai Winichakul

Winkelbauer, Robert, 167

Winkler, Erwin, map of German settlement in Czechoslovakia, 151

Winner, Langdon, and politics of technology, 96, 193n15

Wood, Denis, 147

World War I: conditions for armistice, 146; and national self-determination, 145–46; and reduction of German territory, 146. *See also* German map campaign, during Weimar Republic

Wüst, Wolfgang, 165

Yayci, Cihat, 84

Yaycıoğlu, Ali, 8, 99

Yiwu County Gazetteer of 1640, 51–58, *53, 59*, 186n13, 186n16; and elements on map of Chongde township, *54*

Yuan dynasty, 56

Zhang Bangchang, and creation of State of Great Chu, 49

Ziegfeld, Arnold Hillen: *Karte des Deutschen Volks- und Kulturbodens* (Map of German ethnic and cultural lands), 155–56, *156*; promotion of maps as tools of persuasion, 155–56